Trauma Industrial Complex

Praise for *The Social Distance Between Us*

'This is McGarvey at his best, asking discomfiting questions of man – most? – of his readers and also pointing out that class inequality is endlessly reproduced by people who either do well out of it or are too institutionalised to see what is in front of them. The quality of McGarvey's reporting and storytelling is first-rate ... he makes no end of astute points.'
John Harris, *Observer* (Book of the Week)

'An Orwell for today's poor ... By the end readers will be left in no doubt about the fact that our society is still riven by class inequality.'
The Times

'Breaks your heart and boils your blood.'
The Big Issue

'Vital and indispensable. Documents how we succeeded in creating a twenty-first-century ruling class who – in their complacency, their lack of engagement, their blinkered ideology and dead-hand managerialism – are themselves, now, the principal source of the social problems they so confidently locate elsewhere, and which they therefore cannot even begin to solve.'
Joyce McMillan, *The Scotsman*

'An essential read for every politician, civil servant, councillor, charity worker, police officer and teacher. [An] angry, but controlled, expose of the wide gap between Britain's decision-makers and those most affected by their thoughtless, stupid or selfish actions.'
Susan Dalgety, *The Scotsman*

Praise for *Poverty Safari*

'Nothing less than an intellectual and spiritual
rehab manual for the progressive left.'
Irvine Welsh

'Another cry of anger from a working class that feels
the pain of a rotten, failing system. Its value lies in the
strength it will add to the movement for change.'
Ken Loach

'*Poverty Safari* is an important and powerful book.'
Nicola Sturgeon

'*Poverty Safari* documents in vivid, piercing and frequently
funny prose, the reality of growing up in Pollok and
the consequences of a chaotic family life.'
Stephen McGinty, *The Sunday Times*

'By his own account, Darren McGarvey's first twenty-five years were
a real-life version of *Trainspotting* ... *Poverty Safari* [is] a painfully
honest autobiographical study of deprivation and how society should
deal with it ... But what has made McGarvey such a particular
figure of attention is his political message ... [McGarvey] seems to
offer an antidote to populist anger that transcends left and right
... his urgently written, articulate and emotional book is a bracing
contribution to the debate about how to fix our broken politics.'
Financial Times

'*Poverty Safari* is one of the best accounts of working-class life
I have read. McGarvey is a rarity: a working-class writer who has
fought to make the middle-class world hear what he has to say.'
Nick Cohen, *Guardian*

'Raw, powerful and challenging.'
Kezia Dugdale

Also by Darren McGarvey

The Social Distance Between Us (Ebury Press)
Poverty Safari (Picador)

DARREN McGARVEY

Trauma Industrial Complex

*How Oversharing
Became a Product in
a Digital World*

EBURY
PRESS

EBURY PRESS

UK | USA | Canada | Ireland | Australia
India | New Zealand | South Africa

Ebury Press is part of the Penguin Random House group of companies
whose addresses can be found at global.penguinrandomhouse.com

Penguin Random House UK
One Embassy Gardens, 8 Viaduct Gardens,
London SW11 7BW

penguin.co.uk
global.penguinrandomhouse.com

Penguin
Random House
UK

First published by Ebury Press in 2025

1

Typeset by Francisca Monteiro

Printed and bound in Great Britain by Clays Ltd, Elcograf S.p.A.

The authorised representative in the EEA is
Penguin Random House Ireland, Morrison Chambers,
32 Nassau Street, Dublin D02 YH68

A CIP catalogue record for this book is available from the British Library

ISBN 9781529103892

MIX
Paper | Supporting
responsible forestry
FSC® C018179
FSC
www.fsc.org

Penguin Random House is committed to
a sustainable future for our business, our readers
and our planet. This book is made from
Forest Stewardship Council® certified paper.

Contents

CONTENTS

Preface

There's a contradiction at the heart of this book, which I must acknowledge up front. In writing about the culture of public storytelling around trauma and the dangers of oversharing, I inevitably divulge parts of my own personal story. In warning against the rush to public declaration, I unearth revelations of my own. This paradox has followed me throughout the writing process. If I argue that we need to think more critically about when, how, and why we tell our stories, what does it mean for me to set mine down on these pages? Having given it deep consideration, as well as seeking advice and guidance, I've reached the conclusion that the only way to make these arguments convincingly is to subject myself to the same standards and scrutiny I'd apply to anyone else who shares their experiences in a public way. This book isn't just analysis of the ongoing cultural conversation around trauma, or the rise of lived experience advocacy – it attempts to embody those principles in practice. I become the case study, drawing the reader into the very narratives I also attempt to deconstruct. But unlike many personal stories of trauma told by survivors themselves, I don't just have experience of sharing my adversity in a public way – I have experience of becoming a public figure who is repeatedly identified and framed by it. This book allows me to apply everything I've learned from being a public face of issues like poverty, addiction and trauma – issues of which I have first-hand experience – but instead of my story being extracted by a broadcaster or charity with their own agenda, well-meaning or otherwise, here I take complete control of my narrative. In doing so, I hope to highlight that while all personal experiences matter, and that anyone who shares them earnestly in the public square is undeniably brave, there is (or should be) a structured process to publicly airing adversity where duty of care is paramount to every other concern – a process I fear we have lost sight of in the race to be seen to 'platform' the 'voiceless'. This is not an argument for silence, of course. It's not a rejection of storytelling, nor a criticism of those who have found purpose or healing in sharing their pain. I have done it myself. I know the relief that comes with feeling heard and validated as well as the joy of learning how stories of our difficulties have inspired others to keep moving or stick around. But I also know the costs of becoming publicly identified with those difficulties. I have learned – sometimes painfully – that sharing doesn't always bring clarity, that broadcasting trauma doesn't necessarily lead to healing, and that the impact of telling one's story is not always within the control of the storyteller. I wrote this book because I have lived through the cycle of disclosure and its aftermath. I've seen trauma storytelling become a commodity, a currency in advocacy, media, and activism. I've seen personal pain shaped, edited and packaged for consumption, often serving others more than the storyteller themselves. I've seen how public vulnerability can trap people in fixed identities, making it harder to grow beyond the story they've told.

I don't claim to have the final word on this subject, but I do believe the conversation is necessary and that I am suitably placed to initiate it. My position in this conversation around trauma is unusual – I have a public platform, my work is recognised in fields where discussion and debate around trauma dominate, but unlike most people who tell their stories or, indeed, those organisations and enterprises that facilitate them, I don't draw funding or a salary from an organisation or institution. I don't have a boss and therefore enjoy the rare privilege and occasional burden of speaking freely while still making a living. I believe this means I have a responsibility to say what so many have been discussing privately for years. If we truly care about the wellbeing of those living with trauma, we must reconsider the wisdom of modelling a process that encourages impulsive, unstructured public exhibitionism. Disclosing our truths is an essential part of healing – but doing so publicly is not. This book is an invitation to think more critically about the culture of storytelling around trauma, to question how stories are elicited and consumed, and to reflect on what we owe ourselves and each other in the telling. So yes, there is a contradiction here, but one I accept now as unavoidable. Because to interrogate the nature of personal storytelling, one must tell a personal story. And this is mine.

Introduction

Trauma has become one of the most dominant cultural issues
of our time. It is everywhere, it seems – shaping our institutions, politics and even the language we use to understand
ourselves and our relationships. It has redefined how we tell
our stories, how we seek validation, and how we make sense
of suffering. But has our current way of engaging with trauma
genuinely served us? Or has it also, in many cases, become
something else entirely – a currency, a performance, a mechanism that traps people within the very pain they seek to escape?
This book is not just an analysis of the ongoing conversation
around trauma, nor is it simply a testimony of my traumatic
experiences. It is an attempt at a commentary on what it means
to be a 'person with lived experience of trauma' in a culture
that increasingly demands ever more personal testimony from
survivors. It examines how trauma narratives are shaped, packaged and consumed – how they are extracted for advocacy,
activism and media, often benefiting institutions more than the
people whose pain forms the basis of the story. It also con-
siders the transactional nature of storytelling in this context.
If personal accounts of trauma (or any other form of adversity
for that matter) are to be more than just a way to lend authenticity to third parties eliciting it, then we must ask: how can it

be used to affect real change rather than reinforcing the same power dynamics it seeks to challenge?

The book's structure mirrors a healing journey. It does not remain static in form but shifts as the reader moves through it, evolving alongside the argument. It begins in the familiar structure of non-fiction: presenting a thesis, defining key concepts and offering illustrative examples to build its case. But as the book progresses, so does it structure – echoing the non-linear nature of recovery. The analysis gives way to an immersive personal story, pulling the reader into the mechanics of autobiographical storytelling itself – how these narratives move, how they manipulate, and how they reflect the way the storyteller wishes to be seen. By the final section, the structure shifts again, taking a more critical view of my personal narrative, forcing me to hold myself accountable in a way few ever would in public. In an era of affirm-only approaches, where questioning dominant trauma narratives is often seen as invalidating them, this section may be uncomfortable. But if the insistence is that all stories matter and deserve to be heard, then surely that must include perspectives that challenge the prevailing orthodoxy. The first part of the book lays the groundwork. It explores trauma's recent salience as a topic, and why personal testimony has been elevated as a form of cultural capital. It questions whether the pressure to share our wounds serves us – or if, in some cases, it exploits us. It asks whether we have lost sight of the risks that come with publicly sharing while examining the power of storytelling, both as a means of healing and as a force that, when wielded recklessly, can cause harm.

Part 2 turns this on its head by advancing the argument through a highly personal narrative – my narrative. As with all personal narratives, its fragmented and occasionally contradictory nature attempts to capture the role stories play in helping us make sense of chaos, but also how they may sacrifice nuance and complexity. If trauma shapes our worldview, it also shapes the stories we tell about ourselves. But are these stories always useful or even true? May they occasionally become self-fulfilling, reinforcing identities rooted in pain? This section explores whether public declaration of trauma, despite its intentions, sometimes achieves the opposite of what it claims – keeping people trapped in cycles of suffering rather than offering a path beyond them. It examines the toll of public visibility on those who share their traumatic experiences more widely than the average person, the expectations placed upon them as a result, and the pressures of performing pain for public consumption.

The final part of the book is about release. If suffering has become a defining part of who we are, how do we begin to move beyond it? Can we rewrite our story without feeling like we are betraying ourselves or the past? This section challenges the notion that our stories are our sole property, examines the role of identity (another hot topic) in keeping us locked in personas riddled with trauma, and whether over-identifying with victimhood is empowering or another form of psychological entrapment – a trauma response. It also asks the larger, more uncomfortable question: if trauma is in part a byproduct of systemic injustice, who is responsible for healing? In an age where political dysfunction, institutional neglect and a winner-takes-all culture war, where untreated trauma and victimhood so often bend the light of truth, how should the most vulnerable, left to fend for themselves, respond?

This book is not just for those immersed in discussion and debate about trauma. Nor is it just for activists, campaigners or professionals who navigate the politics of lived experience for a living. It is for anyone who has ever felt, even vaguely, that the story they tell themselves about who they are and why might be incomplete. It is for those who have a nagging sense that something doesn't quite add up. At its core, this is a book about mental health. But not in the way we have come to expect. It does not offer platitudes about self-care or vague affirmations of self-worth. It is not a guide to healing or a manifesto for resilience. Instead, it is an invitation to be radically honest with ourselves – to hold our own narratives up to the light and ask: Is this story true? Not, is it comforting? Not, does it justify my pain? Not, does it make sense to others? But is it real? We live in an age where there is a great temptation to conceal the truths of who we are behind cultivated identities and political rhetoric. An age where trauma can be used as both a shield and a battering ram, politically. We align ourselves with movements, ideologies and communities that reflect back to us the parts of ourselves we find acceptable, editing out the contradictions, the uncertainties, the darker edges we'd rather not acknowledge. We build identities not just on what we believe, but on how we wish to be seen. And yet, for all the validation this performance brings, so many of us feel unmoored, increasingly disconnected from the core of ourselves. This book is not here to tell you what your story should be. It is not here to dictate how you should process pain or define your identity. What it does offer is a model for re-examining the narratives we cling to – especially those we have outgrown. It shows how stories shape our circumstances, our relationships, and even our sense of possibility. It asks whether we are truly living according to our own understanding of who we are, or if we have outsourced that understanding to the expectations of others. For those who have already sensed the disconnect, who feel that their current narrative no longer serves them, this book models how to perform a page-one rewrite. It is not an easy process. Letting go of an identity – especially one that has protected us, explained us, given us a sense of belonging – can feel like annihilation. But it is often the first step toward something more real, more expansive, and more deeply ours. If you have ever felt trapped by your own story, if you have ever questioned whether the way you present yourself aligns with what you know to be true, then this book is for you. Not to give you answers, but to help you ask the right questions.

PART I

Establishing Safety and Building Awareness

'The best minds in mental health aren't the docs. They're the trauma survivors who have had to figure out how to stay alive for years with virtually no help. Wanna learn how to psychologically survive under unfathomable stress? Talk to abuse survivors.'

Dr Glenn Patrick Doyle (2020)

A Concept in Flux

What is trauma and where
has the conversation gone wrong?

A low winter sun struggles against the back of a pull-down blind, its glow a stark contrast to the cold grey hue washing the front room. As a broadcaster, interviewing people who've experienced great adversity, in the hope that their stories may humanise serious social issues like poverty, addiction or violence, is easily the most challenging part of the job. Today – about to begin filming such an interview for my BBC Two documentary film series, *The State We're In* – is no exception. Across a coffee table in an Airbnb this brisk January morning in Merseyside sits a broken woman with an experience to share – the worst experience of her life. 'Ava was an outgoing girl,' Leeann White tells me, her voice already cracking. As we begin our slow, deliberate descent into her story of losing her daughter to a senseless act of violence, Leeann's face tells a tale of its own. Before the next question passes my lips, I hesitate. While the specifics are paramount, and the story must be told in all its heart-wrenching detail, all I wish to obtain on the lower slopes of this climb is her trust. This isn't the first time she's told her story, and it won't be the last. I'm one of countless broadcasters she's encountered over the years since her life was torn apart. By now she'll be thoroughly aware of the tricks of our trade and how to spot an ambulance-chasing hack a mile off. Aware her experiences with media will be varied in quality over the years, I'm keen to signal that today she's in safe hands. As the obligatory small talk fades, any momentary silence between us seems to stretch for an eternity, and my next question feels like a careful step along a crumbling cliff edge. 'Ava wanted to go travelling,' she says, her eyes lighting the room for the briefest moment. 'She wanted to be an air hostess and planned to move to LA.' As we tread through the tragedy that has defined her life since 25 November 2021, I remain mindful of the unstable terrain. That evening, 12-year-old Ava White left her home to watch the Christmas lights with friends in Liverpool. Like any mother, Leeann was nervous but knew she couldn't keep Ava at home for ever. 'I'd just got home from work when my sister called,' Leeann recalls. Her expression shifts from sombre to bewildered as she relives the moment. '"Ava's been stabbed," she told me. I can remember taking the phone away from my ear, then putting it back and saying, "You've made some kind of mistake." But she hadn't. She said, "No, you'd better get here now."' This is the moment Leeann's old life as a doting mother of two young girls ceased to be. It's the moment she found herself trapped in a purgatory between disbelief her daughter had been hurt and absolute certainty the situation was grave. This is the moment – the point of rupture – when the trauma, which Leeann has lived with every day since, began taking hold of her nervous system. She recounts the sequence of events leading up to Ava's murder. Events I'm sure she's turned over in her mind thousands of times. Ava and her friends were near the Royal Court Theatre in Liverpool city centre when a 14-year-old boy, accompanied by a group of friends, began recording them on his phone. Ava asked him to stop and delete the recording. A standoff ensued, with verbal barbs exchanged. The girls, unarmed, ran toward the group of boys to deter further harassment. In the chaos that followed, the boy produced a flick-knife and stabbed Ava in the neck. He and his friends rapidly fled the scene, later attempting to dispose of the murder weapon. 'Don't leave me,' Ava reportedly pleaded with her friends as she lay fatally wounded, awaiting an ambulance. She was rushed to the hospital, but despite doctors' best efforts, was pronounced dead at 10.16pm. As Leeann speaks, I lose track of the details, drawn instead to her face. Her expression, a hypnotic blend of rage and sadness. In her eyes, I see the same desolation parents display when cradling their dead children in images beamed from war zones. Leeann appears haunted not only by the tragic death of her child, but also by the reality she lives in a world where such cruelty is even a possibility. To be so violently disabused of the illusion that your children are safe would surely leave an indelible mark. It certainly has on Leeann. Despite the odds of such an event happening twice to one family, she gives the impression of a woman preparing for the worst. To my eyes, on this day, her peace of mind appears unrecoverable. Three years after Ava's death, she remains psychologically tethered to the event. There are brief moments throughout our conversation, when Leeann's nervous system seems to glitch, and she presents as though the tragedy occurred just hours ago. The impact on everyone connected to Ava has been catastrophic. 'Our family's been ripped apart,' she says. 'We were always a close family.' Ava's older sister remains by Leeann's side, along with her sister June. But the once tight-knit family structure now lies in ruins. In her grief, Leeann has turned inward and away from many of her loved ones, including the kids. 'I've got six sisters. We were always together. We can't be like that anymore. I can't have my nieces or nephews coming over because they're all around Ava's age. We don't spend Christmas together. I can't do anything anymore. I just like to be on my own.' Leeann describes her retreat from her wider support network as if it were a choice she made herself when the truth may be that trauma has simply taken the wheel of her life; under the duress of its influence, nurturing relationships demand more than her nervous system will currently allow. Her trauma, activated naturally, appears to me unprocessed, effectively taking Leeann hostage in her own life. In her grief, she has adopted life-limiting beliefs about her capabilities which now manifest in conditions of greater isolation. This relational retreat likely feels to Leeann like a wise decision, comforting even, when in truth she may be creating conditions that prolong her suffering. As she shows me photographs of Ava taken on brighter days than this, I wonder how many times she has told her story and whether recounting it so readily hinders as much as it helps. Unlike many contributors I've had the honour of interviewing over the years for my documentary films, who have told their stories so often they develop a rehearsed rhythm, Leeann's story feels raw, and agonisingly current. Like many who suffer such tragedy, Leeann hopes sharing her experience will help build awareness of the impact of knife crime and secure funding for blood-packs to be installed locally which may save lives in future. She is, however, clearly most animated by the prospect of a change in the law which would see harsher sentences for possession of knives. And why shouldn't she be? The intense desire, bordering on compulsion, to advocate for change that might prevent others from suffering a similarly cruel fate, is a common response to a tragic event – the inciting incident that spurs many into lifelong campaigning. Stories like Leeann's keep issues like knife crime prominent enough in the public mind, that they demand engagement from decision-makers and political leaders. The emotional weight of Ava's shocking death, in this context, becomes the means by which wider awareness is raised and political pressure for meaningful change is applied. But an immense toll can be exacted from those who rise to any level of prominence as a result of telling their story of life-altering tragedy. As the conversation draws to a natural close, and I ponder whether I have conducted my interview in the safe and respectful manner I'd hoped, my heart breaks for Leeann and her family. The death of a child is every parent's worst fear. For those who face this reality, the pain is incomprehensible. Leeann must endure a burden far beyond grief. She must walk among us – those for whom such loss is merely a fleeting thought – knowing that even the most empathetic listener could never fully grasp her anguish. To carry such a wound can be a solitary existence. Like armed service personnel who struggle to reintegrate into their community due to the scars they carry from war, parents who lose children may begin to feel like outsiders due to the severity of their pain – and the fact they see everyone else still going about their business unaware of just how cruel this world can be. For those untouched by such trauma, Leeann's heartbreaking account of losing Ava, given so freely, is a wonderful gift, providing a rare and vital window into an unthinkable reality. The final mentions of Ava's name bring tears once again to her tired, bloodshot eyes, pulling some dislocated part of her back into the nightmare. For viewers of the documentary film in which Leeann will be featured, her story will be framed as one of a decrepit criminal justice system, failing working-class communities and personal bravery in the face of unspeakable grief. But the story her eyes tell me throughout our encounter is something else entirely. In all my years of working on the front line as a broadcaster, journalist and community practitioner, Leeann White's is the most severe case of untreated trauma I've ever encountered.

*

Trauma is the imprint of overwhelming distress – an event, series of events or relentless strain that shatters a person's sense of safety, leaving behind a raw and open wound. It lingers in the body, rewires the brain and reshapes perception, distorting past, present and future in its wake. This working definition won't satisfy everyone but sets the table for our discussion by underscoring a central challenge in any conversation about trauma: even among professionals, the term is inherently unstable. In contemporary culture, 'trauma' carries multiple meanings, ranging from a clinical descriptor of profound psychological distress to a colloquial shorthand for minor inconveniences – such as a bad commute, an unsettling film, or a contentious online exchange. Hollywood actors speak of the 'trauma' of bad reviews, illustrating how the term has expanded far beyond its original context. Historically, 'trauma' comes from the Greek word for wound and initially just described physical injuries. It wasn't until the late nineteenth century that its meaning expanded to include psychological symptoms, as psychotherapy began to take shape. The industrial violence of the twentieth century, marked by world wars, the Holocaust, and conflicts in Vietnam and the Gulf, highlighted the psychological scars left on survivors. Media advances normalised trauma-related terms like 'shell shock', and in 1980, PTSD (post-traumatic stress disorder) was officially recognised by the American Psychiatric Association. Parallel to widening clinical investigation, public interest in trauma has also intensified. This interest, however, is less about understanding war veterans or genocide survivors and more about individuals seeking explanations for their own feelings of insecurity, exhaustion and relational difficulties. Many hope that naming their pain as 'trauma' will bring clarity and relief. For some, it undoubtedly does. However, others are discovering trauma is a little more complicated than perhaps they assumed.

Just ten years ago, it would have been difficult to get the average person engaged in a conversation about the most basic principles around this concept. Today, trauma is everywhere, and everyone has it. Across the Western world, millions are framing their lives through the lens of trauma – a concept which has undeniable cultural salience. *Vox* magazine described trauma as 'the word of the decade' in a 2022 article which, among other things, pondered whether the term's inevitable semantic journey from clinical label to colloquial mainstay may soon render it meaningless. Trauma shapes public debates on education, criminal justice and healthcare. It's cited as a root cause of addiction, mental health issues and relationship breakdowns. It permeates media – from music and television to films and books like this one. An entire subgenre has emerged, populated daily by content creators discussing their 'trauma bonds', 'trauma responses' and 'trauma-informed approaches'.

A quick search for 'trauma' online reveals countless intersecting frameworks, discussions and debates, all sharing the same fundamental flaw: each treats its definition of trauma – and the solutions flowing from it – as settled science rather than contestable claims. Whichever rabbit hole you descend, the moment you interact with this content, the algorithm floods your feed with more of the same. Meanwhile, the individuals and companies producing it are incentivised by the mechanics of tech platforms to prioritise frequency and style over substance. To remain visible in a hyper-competitive marketplace, where people searching for relief are bombarded by misapplied terminology, faux-medical jargon and relentless calls to action, creators often resort to bite-sized novelty and superficial explanations. As a result, much of the online content becomes entertaining digital comfort-food, masquerading as reliable advice, leaving those seeking answers with little more than the fragmented simulation of therapeutic guidance. This trend deeply concerns me.

Western trauma culture, preoccupied with attachment styles, personal boundaries and individual resilience, often ignores widespread traumatic experiences rooted in systemic inequalities. It demands we focus narrowly on ourselves, our perceived pain, and the 'movies' playing in our minds. The rise of an unregulated online mental health marketplace amid such stark social and economic decline should alarm us all. I now have mixed feelings about the topic's rapid popularisation, despite initial enthusiasm. Even Bessel van der Kolk, whose book *The Body Keeps the Score* became a go-to bible for the trauma-curious during the Covid pandemic, has voiced concerns about the increasingly broad application of the term. In a June 2024 interview with *Time* magazine, van der Kolk remarked:

'People are inflating the whole trauma notion and now apply it to everything. When somebody breaks up with you in a love relationship, that's part of life, but that is not a trauma.' Naturally, van der Kolk also has his fair share of detractors who dispute his conclusions, dismiss his fanbase and detest his influence. What cannot be disputed is that trauma is the latest term to suffer the dreaded 'concept creep'. Concept creep refers to the gradual broadening of the definitions of concepts – particularly those related to harm, mental health and social issues – to include a wider range of phenomena over time. This term, coined by psychologist Nick Haslam, describes how ideas like 'trauma', 'abuse', 'bullying' and 'addiction' have expanded to encompass milder, less severe or different experiences than they historically did. It is of course a feature of progress that definitions and labels change, and terms are expanded to encompass new developments. And it's also the case that those of a more conservative disposition may resist this as a matter of instinct – not always for sound reasons. However, the rapid commercialisation of trauma – powered by social media influencers and online therapy – is a fairly novel and radical force. A marketplace that is reshaping the mental health space faster than we can safely regulate it, rife with perverse incentives much like those found in private healthcare systems, should alarm all of us. In these highly competitive commercial conditions, trauma is too often reduced to bite-sized maxims and mental health minutiae. While the increasing openness about trauma is welcome, it has been accompanied by a parallel explosion of disinformation. This commodification of trauma, which is now as much about identity as it is about health, empowers individuals to self-diagnose or diagnose others without oversight. It encourages the drawing of conclusions as to how trauma may co-occur with other conditions like ADHD and personality disorders. And it welcomes us to view almost every aspect of our lives, from family dynamics to romantic intimacy and even our political beliefs, through a trauma lens. Central to this poorly lit hall of smoke and mirrors is the powerful and misguided notion that telling your story in a public way is an essential part of any healing journey.

The Trauma Industrial Complex is the system in which the very real and urgent issue of mass unacknowledged, untreated trauma is commodified, medicalised and repackaged for profit, validation and influence. It extends beyond clinical care into self-help industries, social media, civic society and corporate wellness, where trauma narratives are amplified, exploited and even monetised. Genuine suffering is often flattened into marketable content, while trauma discourse expands to encompass everyday struggles, blurring the line between distress and disorder. As public services collapse, individuals turn to online diagnoses, influencers and expensive pseudo-therapies, creating a cycle where trauma becomes both a personal brand and an economic driver, rather than a wound to be meaningfully healed. In my view, the most profound effect of this eruption in trauma-related discussion is the powerful, but deeply misguided idea it often models, that publicly sharing pain is how you recover. People with experience of great adversity will increasingly put themselves forward to disclose the painful details of their wounds in the public domain. And often, they will be invited to do this before it has been determined – by either the storyteller or the facilitator of the story – whether broadcasting it in a rowdy and unforgiving public square is either useful or safe. While there is much to dissect in this brave new world of trauma, this book will focus mainly on that particular facet of this Trauma Industrial Complex – the notion of cathartically sharing one's 'story' – the consequences of which remain inadequately explored.

Many of you will be familiar with the ongoing conversation about trauma I describe and are likely no strangers to other forms of media that centre first-hand accounts of adversity. In recent years we've seen a rise in gritty depictions of trauma in television shows and memoirs, but where is the narrative about the toll this public exhibitionism takes on the storyteller? That's the story I'm here to tell. Culture is awash with personal testimony of trauma, but one story we haven't heard in any great detail yet – perhaps the only one – is the story about what it's like to be that person, whose experience is sought-after by media, charities, decision-makers and consumers. I'm one of those people – a so-called lived experience campaigner. My story of childhood trauma, poverty and addiction is really rather ordinary where I come from, though the way I tell it isn't. For much of my life, I have used that story to highlight the plight of those who have suffered trauma caused by societal dysfunction. I've shared my experiences to capture the attention of decision-makers. And I've upcycled it to create my best attempts at art. But one day, my story escaped the confines of conference halls and community centres and began appearing in the national press, eventually making its way around the globe, translated into numerous languages. While this is the dream for any writer, and the aspiration for many creatives and campaigners, there came unforeseen consequences of this prominence. Consequences I'm certain many others in my line of work have experienced, though perhaps feel apprehensive to disclose. It is therefore my sincere hope, that as well as offering a brief, albeit unacademic critique of the Trauma Industrial Complex, and guidance on navigating the rapacious cultural demand for highly personal, autobiographical storytelling, this book will also attempt to tell the story of how I – someone who became internationally noted for publicly broadcasting my adversities – might do some things differently if I could go back.

'The process of constructing a narrative of the trauma is not only a means of integrating it into one's life, but of regaining control, reestablishing the self that was shattered.'

Susan Brison, *Aftermath: Violence and the Remaking of a Self* (2002)

CHAPTER 2

The Rise of
Catharsis Culture

*How did trauma narratives come
to dominate and why are we so compelled
to share our wounds in public?*

My journey as a poverty poster-boy began in 2001. Like many
who find themselves in this unlikely position, it was not something I had planned – I was too busy surviving. The sudden
death of my alcoholic mother aggravated a long-running rupture in the family household which left me estranged from
relatives and within a few years, homeless and on the brink
of alcoholism myself. Hip-hop music became my outlet, and
I began performing locally under the name Loki – a moniker
borrowed from the shape-shifting, home-wrecking Norse god
of mischief that I first discovered in the film *Dogma*. What
felt like a world away from the housing estate I grew up in,
open-mic nights and rap battles became my new stomping
grounds. Using autobiographical rhymes to vent my adolescent fury over dusty boom-bap drumbeats, I forged a whole
new persona that many who knew me from before would have
found unrecognisable had they encountered me. Like many in

my generation – a reluctant millennial – I came of age at the dawn of the twenty-first century, a time of digital revolution. In this rare cultural sweet spot, information was becoming more accessible, yet the internet hadn't fully infiltrated our lives. It was a simpler, less-informed era, where traditional media had the final say on what was true, who was good or bad, and what was considered cool. The family doctor was still the authority on what ailed you, and pop stars, not podcasts, ruled the airwaves. Back then, the biggest artists sang, shouted and rapped about their personal struggles. Unlike the cryptic angst of Nirvana – where fans speculated on the meaning behind songs like 'Something in the Way' – the pop stars of my day were far more direct and literal. Queens gangster rapper 50 Cent made his name recounting a tale of being shot nine times. Kanye West climbed the charts with his story of surviving a near-fatal car crash that left his jaw wired shut. The most impactful artist by quite some distance, however, was Eminem, whose raw accounts of trailer-park poverty, familial dysfunction and child neglect at the hands of his pill-popping, alcoholic mother dominated airwaves and headsets worldwide. Rightly criticised for his rampant misogyny and homophobia, Eminem also spoke very directly to adolescent males like me who had been let down badly by their mothers – a finer point many commentators of the day missed when attempting to contextualise his popularity. While distinct from every other charting artist of the day, Eminem mainstreamed a trend in music which is now commonplace: publicly airing trauma no matter the cost. These artists weren't trying to change the world like Public Enemy or wax philosophical about modernity like Radiohead. Instead, they focused on themselves, offering shockingly detailed accounts of their adversities. By venting their frustrations, heartaches and traumas in their music, they modelled a form of self-expression for troubled teens like me, who were searching desperately for identity. This confessional style of storytelling was part of a broader cultural shift. The 2000s were a golden age of airing dirty laundry in public, and not all of it was high art. From bear-baiting daytime talk shows to prime-time talent competitions, people routinely risked humiliation for a shot at fame and fortune – for the chance to feel loved, accepted and secure. Everyone, it seemed, was looking for a ticket out of life on the estate and I was no different. My reputation grew in Glasgow's music scene, and I eventually became a community artist. When the third sector discovered me – and my 'story' – I found myself in high demand, sometimes taking the stage at conferences several times a year. Professionals sat slack jawed as I shared my experiences. Whether campaigning for Scottish Independence, criticising decision-makers for failing the most vulnerable or releasing new music delving deeper into my trauma, my work seemed to resonate more when anchored in my woundedness and willingness to self-reflect. This growing public interest culminated in the publication of my debut book, *Poverty Safari*, in 2017 – a mix of memoir and social commentary in which I revealed some of the traumatic events of my childhood, quite unaware of just how many people would go on to read about them. That book changed my life, and I never looked back. I had overcome the odds, finally redeemed as a figure people admired. They loved my story – they loved me.

In the past two decades, this particular type of personal storytelling has seeped into every corner of culture. Charities, politicians and philanthropists have embraced lived experience as the gold-standard perspective – the latest silver bullet.

Rooted in feminist standpoint theory, the idea is that marginalised voices provide unique insights into social realities. It's an argument I've made myself. It even formed the thesis of my last book, *The Social Distance Between Us*, in which I argued that a lack of proximity between decision-makers and the lived realities of those their decisions impact, lies at the root of the UK's political and economic breakdown. My conclusion was that in order to bring balance to society, historic class inequalities in education, health and the labour market must be addressed, and such a process would be best served if real-world experiences were included in the policy design process. Consider, as a simple example, the premium-rate customer helplines previously used by the Department of Work and Pensions, whereby the poorest in society were expected to pay for the privilege of corresponding with the government agency set up to support them in economic difficulty or illness. Would such a ludicrous situation have arisen had people with experience of benefit dependency been consulted beforehand? I have my doubts. In this context, first-hand knowledge of what it's like to rely on benefits to survive is immensely useful, if the aim is to genuinely lift people out of poverty rather than punish them for being poor. The usefulness of including lived realities in social policy formation is self-evident and one of the simplest, most effective ways of eliciting those experiences is by inviting those affected to share their stories. In recent years, however, these stories have become increasingly untethered from their original purpose. With the onset of social media and the rampant capitalist individualism which commands us to view ourselves as products and our experiences and beliefs as brand extensions, standpoint theory often becomes depoliticised from any discernible aim, and trauma decontextualised from the material realities which make it likelier to occur. Stories of trauma are now invariably told for their own sake, driving less at the notion of systemic change. While this book primarily focuses on the role of stories in trauma, it's worth noting that individuated narratives, while still serving a cultural purpose – providing catharsis and identification, particularly among minority groups – have become largely dislocated from their original function – to bring about a more just society for all. A story's prominence often hinges on its mass appeal, and nothing is more divisive in today's climate than an overtly politicised tale of adversity. The masses will tolerate, and even enjoy, the blood, guts and tears of trauma – most people are on board for those – but dare attribute social suffering to a political decision, party or figure, and your campaign can say a sweet goodbye to any hope of widespread public support. This creates incentive for organisations who rely on such stories to depoliticise them, and in the process, dial up their sentimentality or shock factor. In the process, this models a manner of storytelling to others, who come to believe the mere broadcast of their personal pain may be enough to change the world. Stories are told for all sorts of reasons, but stories of trauma take a particular shape, possessed of a certain emotional signature, which often sacrifices truth's complexity in pursuit of wider appeal. This, in my view, is a problem, and the time to address it is now.

*

Our fascination with the trials and tribulations of others is not hard to fathom – we are nosy buggers by nature, and we all love a good story. Even if you don't consider yourself traumatised, you likely have a little movie playing in your head about your own life, too. Stories are how we make sense of the world and ourselves. Every day, we are bombarded with stories in television, music, social media and advertising. Stories of 'plenty of fish in the sea' help us cope with heartache. Stories of the afterlife ease grief and the fear of death. Stories of 'hard work paying off' get us through soul-crushing jobs. The best stories stick with us. Yet the stories we know best are the ones we tell ourselves. I've been a storyteller for as long as I can remember. As a five-year-old, I spun a tale to explain how I managed to slam a drawer shut on my own penis while searching for clean underwear after soiling the previous pair. Even then, the truth seemed insufficient – the injury demanded a more elaborate rendering. Over time, my stories evolved – the whoppers of adolescence, the self-serving narratives of heartbreak, and the idealistic tales of youth rejecting societal conventions. Eventually, adulthood forces us to revise our narratives. Youthful tales of romance, adventure and success give way to stories that explain or justify the lives we now live. For most of us, these personal narratives are shared sparingly and safely if at all. But confiding in a trusted friend over coffee is vastly different from disclosing the most traumatic event of your life on national television – something I've done, and I am not the only one. 'A worry shared is a worry halved,' goes the adage. But where does the other half of the worry go? The urge to disclose our troubles is part of what makes us human. When we have to get something off our chest, we turn to friends, partners, colleagues and even counsellors and experience relief or gain perspective as a result. In my own recovery, hearing stories of how other people got well was pivotal in my decision to accept I had a problem and to give sobriety an honest try. Thanks to their openness and vulnerability, I became willing to accept the true extent of my alcoholic condition. It was a truth I eventually conceded in recovery meetings where sharing personal experiences openly is not only commonplace but also deeply practical. In the context of the Trauma Industrial Complex, however, this culture of sharing is not always a two-way street where upon hearing our side of a story, a trusted friend or partner may offer their take, potentially creating greater understanding or insight. Instead, catharsis culture invites us to transmit our stories much like a broadcast, and like television, interactivity is limited. In recovery, we have a saying that goes something like, 'take the cotton wool out of your ears and put it in your mouth', which is a form of tough love imploring us to recognise that alone, we can never find all the answers to our problems. It urges us to open up to the possibility that despite our strongest reservations, we may benefit from listening more rather than insisting on ourselves. Ultimately, if we are only on transmit, taking few to no incoming calls, then what are our stories really worth, beyond fleeting catharsis? For a story to possess real weight and depth, it must be subjected to the light of the truth and told for a greater purpose than the mere satisfaction of hearing ourselves say it. Social media promotes a form of dialogue which is no dialogue at all – it's just us talking to ourselves – and this trend is pervasive in a lot of discussion about trauma.

Daily, we tell stories about ourselves – online, on social media, in conversations, or even in our own minds. We feel a sense of validation when our stories resonate with others. While storytelling is natural, the platforms we now use to share are not. Before social media, our darkest secrets rarely ventured beyond a trusted confidant. Today, we can broadcast the most intimate details of our lives with the push of a button. Even the most reserved among us often maintain multiple social media accounts, effectively becoming our own publicists. A new culture of openness about trauma has emerged, creating a demand for content that spotlights victims and survivors who 'overcame the odds'. Lived realties, once marginalised, are now prominently featured. This represents progress, but it also poses risks. At the heart of this cultural shift are individuals who genuinely believe that publicly airing their pain will help them make sense of their adversities, inspire change or comfort others. Yet, many ultimately find that the validation they seek is fleeting. As the fanfare fades, the underlying trauma – and its companions: fear, paranoia and shame – inevitably resurfaces.

Too often, those who support trauma survivors to share their stories fail to grasp how vulnerable some of these individuals may be or how their stories may find larger audiences than was ever intended. My story has been featured in newspapers, studied by academics and adapted into books. While it has raised awareness, it has also modelled a problematic approach to trauma in which public exhibitionism takes precedent over safeguarding. Despite overcoming significant challenges, telling my story has come at a personal cost, too – something we will explore later. Over time, I've been forced to reflect on the catharsis culture that elicited my stories and reconsider the motivations behind them.

Today, I live in the shadow of a character I created. Despite my attempts to move beyond the lived experience genre and focus on systemic issues like poverty and inequality, I find myself continually drawn back to the same story. My audience, in many ways, won't let me move on from it. Or, at least, that's how it feels sometimes. Any work I produce (music, live shows, books, television) which is not anchored by my own suffering in some way draws tangibly less interest from both audiences and, in some cases, commissioners – a problem when your whole career is based on you being the product. Nobody is forcing me to put my business out there – I choose to do so and understand the risks – but telling the story, even now, exacts a toll. The adversities I recount may be in the past, but their impact remains part of me. This need for validation – for love and acceptance – has not provided the self-worth I imagined it would. While my story has opened doors, and I'm sure helped many people, it has also constrained me. The narratives I shared at moments of vulnerability, lacking proper insight, were often incomplete or misleading, reframing other chapters of my life in harmful ways. And I'm not alone. Many trauma survivors fall into the trap of oversharing, confusing the temporary high of public validation with the genuine, painful process of healing. Too often, trauma narratives dwell on the details of what happened rather than how recovery was achieved because so many people sharing their stories have no real experience of processing trauma – only experience of sharing their adversity. This imbalance reinforces another problem with much of the current conversation around trauma: it traps us in self-portraits we constructed within the narrow confines of our woundedness. Often, the stories we tell reflect what we find easiest to discuss – how we were wronged, who or what was to blame and how trauma manifests in our lives – when what we ought to spend more time considering is why so much of our story is left on the cutting room floor. Trading on a story that touches only part of the truth is to abandon the very 'authenticity' we, the public storytellers, are always assumed to possess in spades. Catharsis culture can leave us stuck in the awareness-building phase of recovery – the first of three – mistaking the safety of retelling our story – often a highly selective draft of the whole truth – for truly getting well. To move forward, we must engage in the hard work of introspection and change, supported by nurturing relationships that are about more than simply affirming and validating us – the odd hard truth goes a long way too sometimes. It's time to shift the focus from simply telling our stories to understanding and addressing the deeper wounds they may represent. There are countless books about trauma and its effects. Few, however, are written by survivors willing to acknowledge certain basic facts: our stories are often incomplete, there may be self-seeking at play beneath our claims that we disclose them publicly only to help others, and ultimately, if either of these facts apply, we may have some work to do before we should be platformed as authorities on important matters like trauma.

'The exposure of pain can be a form
of self-dispossession: in speaking of pain,
you give something away. That gift is
not always received with care.'

Sara Ahmed,
The Cultural Politics of Emotion (2004)

Defining Lived Experience

How our stories inspire, heal and occasionally harm

Kevin Hines was serious about ending his life. So serious, in fact, that his suicide attempt involved a 75-metre jump from the Golden Gate Bridge – the world's most frequented suicide spot. Since 1937, more than 2,000 people have died there, with 98 per cent of jumpers perishing in the icy waters below. Kevin is one of the rare 2 per cent who survive the fall. I first heard about Kevin through a viral YouTube video titled 'I Jumped Off The Golden Gate Bridge'. In it, he recounts in vivid detail what he believed at the time to be his final moments. Kevin is one of more than a million Americans who attempt suicide annually. In 2017, 14 out of every 100,000 Americans died by suicide – a 33 per cent increase since 1999 and the highest rate recorded in the US since the Second World War. Factors such as the opioid crisis, social media and rising stress, inequality and economic instability are often cited as drivers of this surge. But for those living with suicidal thoughts, these societal factors fade into the background. All that registers is the pain.

'So, I was born on drugs and premature,' Kevin says in the video, 'and then I bounced around home-to-home. Nobody wanted to keep me because I was sick.' After years in foster care, Kevin believed his life was turning around when Patrick and Deborah Hines adopted him. 'I had a great childhood. I thought everything's gonna be great. And then at seventeen, it all came crashing down.' Kevin began experiencing severe paranoia that spiralled into mania and hallucinations. Diagnosed with bipolar disorder, he felt his life was 'spiralling out of control'. He vividly recalls writing his suicide note and believing he was a burden to everyone around him. 'That's what my brain told me,' he explains. 'That's how powerful your brain is.' Kevin describes suicidal ideation as a mental state where negative thought patterns become inescapable. What starts as a cloud on the horizon develops into a hurricane that consumes the mind, often compounded by emotional isolation. People may feel a desperate urge to cry out for help but find themselves unable to do so. Problems snowball, perspectives narrow and negative self-talk takes over. Solutions are dismissed as impossible, and the person becomes convinced they are beyond help – a core belief that fuels suicidal despair. Kevin took a bus to the Golden Gate Bridge, where he hoped to end his life. 'People rode by me, drove by me, walked by me,' he recalls, the pain of feeling unseen still evident in his voice. 'The only woman who approached me asked me to take her picture, then walked away. It was at that moment I just said, "nobody cares," when the reality was that everybody cared. I just couldn't see it.' In a split-second decision, Kevin leapt. 'What I'm about to say is exactly what nineteen other Golden Gate Bridge jump survivors have also said: the millisecond my hands left the rail, it was instant regret.' He describes the overwhelming survival instinct that activated as soon as he realised his death was imminent. Within four seconds, he plummeted at 75mph from the equivalent of a 25-storey building into the freezing water below. As he fell, one thought drowned out all others: *Nobody's going to know that I didn't want to die.*

Suicide is a difficult issue to broach for many people and that's partly because they don't know how to safely talk about it. The confusion around suicide is nothing new. For millennia, it has preoccupied philosophers, ethicists and lawmakers. In ancient Athens, those who died by suicide without state approval were buried on the outskirts of the city in unmarked graves. Suicide was criminalised across Western Europe during the Middle Ages and remained a crime in the UK until 1961. Even today, suicide is illegal in many countries and carries severe cultural stigma in others, particularly within the Abrahamic faiths. Despite this long history of punishment and moral condemnation, suicide remains a leading cause of death worldwide. According to the World Health Organization, one person dies by suicide every 40 seconds. In 2017, more people died by suicide than from malaria, homicide, meningitis, terrorism or natural disasters. Yet statistics, however stark, cannot convey the human reality of suicide. Kevin Hines's story does.

Kevin's powerful narrative challenges many pervasive myths about suicide, particularly the notion that those who attempt it truly want to die. While it's not unreasonable to assume that someone who ends their life no longer wants to live, the reality is often more nuanced. For many, death represents not a desire for oblivion but a desperate solution to the unbearable problem of life. Kevin's testimony also underscores that suicidal ideation often results from a mental malfunction – an intrusive, distorted perception of reality. Much like addiction or severe depression, it overrides the brain's natural survival instinct. This distortion explains why suicide rates often defy simple evolutionary or social logic. Hines also debunks the idea that once someone has decided they want to die, they cannot be dissuaded. In truth, he desperately wanted someone to intervene but felt unable to articulate his pain. Those around him – family, friends, and even strangers on the bridge – were unaware of his anguish because he lacked the ability to communicate it. This inability to reach out is one of suicide's cruellest aspects: the help someone most desperately needs feels impossible to seek. This gap is worsened by strained mental health services, which often cannot provide immediate intervention during moments of crisis. Well-meaning platitudes about mental health often fall flat for those in the grip of despair, highlighting the disconnect between public awareness campaigns and the lived realities of those suffering. If we look deeper at the circumstances of Kevin's suicide attempt, we also see there must be an alignment of the desire to end your life with the means by which to do so. Often, people feel they want to die but lack the means to complete suicide in a manner they'd prefer. Conversely, when those means become available, the urge to end it all is not as strong. Suicidal thoughts are far more common than suicides themselves. Often, rather than symbolic of the intent of the departed, completed suicides bear more resemblance to accidents, and like accidents, often occur due to preventable environmental or social factors. In Kevin's case, a suicide net fitted below the bridge would have caught him after his hand left the railing, but due to opposition from certain sections of the community, they were not installed. Thankfully, nets have since been fitted on the Golden Gate

Bridge, ruling the hotspot out for other potential jumpers. These are just some of the lessons we may learn from Kevin's story that even the most resourced and well-meaning public awareness campaigns might miss or clumsily communicate.

Kevin's survival and willingness to share his experience have without doubt saved countless lives, offering a rare and invaluable perspective on suicide. His is a powerful example of the importance of first-hand accounts, which can ripple across communities in profound ways. Inspired by Kevin's story, I wrote a song, 'Don't Jump', which has since resonated with many people in my own community. I was particularly drawn to Kevin's description of instant regret the moment his hand left the railing, feeling it a powerful and original insight you simply would not find in a safe, performatively sensitive mental health awareness campaign. Here, a survivor of a suicide attempt, merely by recounting his story, telegraphs a warning that no matter how a despairing person feels before the attempt, they may spend their final living moments wishing they could take it back. In 'Don't Jump', I lasered in on this idea, certain it would move others out of suicidal ideation in the same way it did me.

Don't Jump,
You'll regret that,
Don't throw it all away,
Take a step back
Staring in that black river
As the waves shimmer,
Doesn't matter if you're a great swimmer
Don't Jump,
It's colder than you think,

As soon as you let go,
You'll be overcome by an instinct to survive,
Too late,
You'll have sealed your fate,
You'll go to your grave
Knowing stepping off was a mistake
Don't Jump,
I can see that you're hurting,
I can see you're suffering,
The rivers lined wae restaurants and casinos
You can't afford to play or eat in
You're told to be grateful to work in,
Pretending you're tougher than you are
To express your masculinity,
No industry left
To feel a sense of purpose or dignity,
While they tell ye that yer problem's privilege,
But think eh yer mates,
Think of your sisters,
Think of yer da's face,
Try and see beyond the mirage insanity paints
On the canvas of yer brain,
As it aches with sadness and pain
Don't Jump,
I know yer sick of the phone
The texts,
Being one of the best and getting no respect,
Losing yourself in a coke sesh,
Going fae sober and fresh to a total mess,
Lost and depressed,
In your own head

Abandoned sleeping bags on shop steps,
Women forced to sell torn flesh
To more men
For the money to score smack off the same guy that's
whoring them
Yer burd won't connect,
And normal porn's a snore fest
Compared to faux-cest and forced sex,
The impulses you develop trying to cope,
Bring you more stress
One more step,
Closer to the edge,
Don't Jump,
Realise yer brilliance,
The mild stimulants
You use to manage hypervigilance,
A toxic algorithm,
Oscillating violently between humility and narcissism,
This city isn't pretty,
Its brutal in its beauty,
But I swear to God,
It would be worse withoot ye,
I know why they dae it,
I know why these boys top themselves,
As I'm standing on this fuckin bridge,
Talking to myself

Hines offers a particular strand of what today is referred to as lived experience. The first-hand account, storytelling strand which many regard as the only kind. As I see it, there are in fact three distinct forms which I'd like to briefly outline before we proceed. Admittedly, the following section may feel a little dry and academic for some of you – not least coming off the heels of a raw testimony like Kevin's – but I urge you to bear with me. As you proceed, I also want you to notice any momentary lapses in concentration or interest, and what these may say about your expectations of people, like me, with first-hand experience of trauma, and the roles we're expected to perform. My experience is that enthusiasm among some readers dips ever so slightly when I'm not drawing directly from my own suffering (or the suffering of others) to illustrate my point. With that in mind, let us continue. The first of the three strands is lived experience surveyed at scale: this is when decision-makers driving at the root of a problem (voluntarily or otherwise compelled by campaigners or events) invite participation from a cross section of people affected by the issue and from the evidence gathered, begin devising solutions. Take an issue like male violence against women as a pertinent example most of you will recognise. Thanks to the testimony of women who've experienced various forms of abuse at the hands of men, a vast body of understanding of the issue of gender-based violence now exists. This body of understanding exists independent of anyone's personal opinions. There are facts about male violence against women we simply know to be true and that remain true in almost every instance it is found. We constructed this body of understanding over time, by listening to women affected. With a large enough sample size, whatever the issue may be, themes will begin to emerge irrespective of individual circumstances or interpretations. Only then do we gain universal insights upon which genuine solutions can be built. We know, for example, that male abuse of women is rarely just physical but also emotional, psychological and sometimes sexual – facts which at one time were minimised, dismissed or contested. We know that just because an abusive male is affable, charismatic and otherwise normal in many other respects – he may be a good father or a well-regarded member of the community – it doesn't mean he can't also be a perpetrator of abuse. And we know most gender-based violence perpetrated by men against women follows a similar if not exact pattern, where male abusers slowly isolate women from friends, family and other sources of potential support, as they tighten their grip on a victim's life. We understand all of this and more after decades of listening to survivors, often platformed by campaigners and advocates, which garnered the public support required to force decision-makers' hands. Evidently, this problem is far from solved – male violence against women remains endemic – but the vital awareness building around it has been largely successful. This is a welcome development we owe to the courage of women with direct experience of male abuse whose individual accounts culminate in a greater understanding of this problem.

Everyone has a so-called lived experience of some kind. Every doctor and nurse has had a health problem, every teacher has attended school, and most cops lived in a community affected by crime. It is often as direct consequence of these experiences that many move into particular roles in their community, feeling those experiences may benefit others. I recently spent a day on the beat with a female cop on her first day in the job. The final call we responded to came from a woman making a plea for help while hiding from a violent male who had broken into her flat through a window. We arrived at the scene, and police entered the premises while my director and I remained outside. We couldn't film, but the officers were wearing microphones, and we could hear the audio. While the seasoned cops dealt with the disturbance, the rookie searched the flat, discovering the terrified woman hiding in her wardrobe. While it would be inappropriate to relay precisely what was said, she comforted the victim with such warmth and empathy, we couldn't believe this was her first day on the job. After she made the arrest, she disclosed to us that as a child, she had witnessed domestic violence perpetrated against her mother and that this experience was partly what drove her into the force. People like that exist across every public service and sector. This is an example of lived experience in social settings – the second form of lived experience. In the addiction field, for example, an increasing section of the workforce is made up of people who have themselves recovered from alcohol or substance misuse. These jobs are not created out of pity. It is understood within the drug sector that addicts often respond more positively to the guidance, suggestions or even criticisms of those who've been through the wringer themselves. Who better to still those excruciating withdrawals in the first days of treatment than someone who's gone through a few rattles themselves? People with first-hand experience operating in specific social settings (health, education, criminal justice) often possess specialist knowledge and intuitions that cannot be taught – they must be lived. While this is not equal to years of academic study, any highly qualified professional worth their salt understands their value in building trust and rapport in community settings. Now and then, such a person may even spot something a highly trained professional would miss. Every GP knows how to write someone a prescription for a mental health problem but how many could spot a drug addict feigning anxiety in the hope of obtaining a safe supply of Valium – I might because I was one. This type of experience can make the difference between a merely adequate public services encounter and an exceptional one. Take two teachers at the same school, equipped with identical resources: one might respond to a traumatised child with love, patience and compassion in response to challenging behaviour, while the other might impose strict boundaries, punishment or even exclusion. The difference often lies in their own childhood experiences and the skills and intuitions they acquired as a result. Lived experience in social settings, much like lived experience at scale, is immensely valuable for reasons which should now seem obvious. What is perhaps even more endearing about this less noted personal pursuit is that countless people bring it humbly to bear every day across society, without any desire or need to sing it from the rooftops. Which brings us to the third and final form of lived experience, as I see it – and to the end of the dry, uninvolving passage I forewarned you about a few moments ago.

While everyone has personal experience and insight of some kind, not everyone feels compelled to share it publicly. That role is reserved for a specific kind of person. As you've probably guessed, I'm one of those people. Kevin Hines is one of those people. Leeann White is one of those people. If you spend enough time online, you're bound to encounter people like us – lived experience storytellers. We often express strong opinions, fuelled by passion and conviction. We believe our experiences matter, that they illuminate the gaps in understanding among a well-meaning but often detached managerial class. From addiction and homelessness to criminal justice, gender-based violence, racism, housing, mental health and trauma, our stories are seen by many – and sometimes by ourselves – as critical pieces of a complex puzzle. We believe that when decision-makers and society truly grasp these lived realities, they can pave the way for a more compassionate and inclusive future.

But that's not the whole story. Unlike the other types of lived experience I've so far outlined, storytellers like me are often highly visible. Our experiences are also commodities, and our stories rely on specific skills and traits that not everyone who's been through such circumstances possesses or develops. Stories are our main currency. They inspire professionals, populate policy documents and fuel funding applications. They may even become bestselling memoirs or world-beating Netflix shows. Every day, stories of adversity saturate culture, driving engagement on social media platforms. Social media posts by grieving relatives are turned into clickbait news. Interpersonal dramas are played out on reality television. Threads, status updates, think-pieces, video essays and news segments are shared, debated and dissected. In this marketplace, the demand for authenticity and social realism feels insatiable. And for those of us supplying it, there's a sense that we're making waves – having an impact, making a difference, maybe even going places.

Storytellers break into roughly two camps: those who are thrust into a public-facing position suddenly by a tragic event and those who, having experienced that public interest in their story, come to leverage it with greater intention, to raise awareness and advocate for change. Those of us who come to advocate for change differ in a key way: we are actively involved in the process of shaping and delivering our stories. Leeann White, for example, is a storyteller, but she is more passive in that pursuit than someone like Kevin Hines – an international public speaker on mental health and suicide.

Leeann may make that transition herself soon, as she continues on her healing journey. My point is simply that some people receive prominence as a result of the shocking details of their stories, while others enjoy platforms because they learn how to shape their experiences and relate them to ongoing social issues. While we may talk like every experience is of equal value, the impact of a story often hinges on the storyteller's charisma, their skill in crafting a compelling narrative and understanding their audience, and the shocking details of their trauma or circumstances. Some stories, like Ava's, are innately compelling because they are so deeply tragic. Others are juiced by the affability or skill of the narrators – I say this with some semblance of self-awareness. For some, talent for storytelling comes naturally; for others, it develops over time.

It may jar with the salient notion that personal testimony of trauma represents the very apex of authenticity, but those of us who've been around the block a few times develop an intuition for shaping our narratives. One essential element of any successful story of adversity is the 'moment of revelation' – the transformative point where the protagonist emerges from the belly of the beast, reborn. This mirrors the classic 'hero's journey', reinforcing two assumptions about trauma that don't always hold true: that recovery occurs after a singular event or intervention, and that having survived the worst, we are now fully healed. Our stories often conform to this structure because it is what we've seen modelled by other storytellers – not necessarily because it reflects reality. Through this process, we unintentionally perpetuate a significant myth about trauma, one Leeann's story contradicts entirely: that recovery is linear and follows a clear, singular breakthrough. Adherence to the narrative may result in playing fast and loose with the truth. The allure of presenting ourselves as 'recovered' – a neatly wrapped, affirming conclusion expected by audiences – can be so powerful that we may begin to believe it ourselves – even when it's not true. Here lies a danger of the narratives we construct for those who may still be vulnerable. As discussed earlier, a cornerstone of the Trauma Industrial Complex is the 'affirm-only' ethos, where survivors or victims must always be believed, and where every feeling is valid. This serves an important purpose in social settings, mitigating the harmful effects of stigma and prejudice while fostering a supportive environment for people to who may have reservations about disclosure. Affirmation also serves a very practical purpose in a therapeutic context as it becomes the basis of trust between a professional and a client. However, in the context of public disclosure, the affirm-only approach can backfire when a storyteller lacks the self-awareness to recognise their own unresolved struggles. For those of us who claim to be more healed than we are – whether knowingly or unknowingly – the pressure to fit our experiences into neat story templates can blur the line between truth and performance. This raises broader questions about the readiness of some advocates to carry a message safely. If we can deceive ourselves in the rush to pour our unprocessed trials into a narrative mould, what other liberties might we unconsciously take with the truth? The reality is that while our voices must be heard, the pressure to conform to pre-existing formulas can sometimes obscure the raw, unpolished truths. The storytelling templates we adopt often strip away nuance, leaving little room for the complex realities. Not all who perpetrate harm are bad or evil. Not all who are victims are saints. Through storytelling, we may inadvertently distort certain facts about ourselves and, in turn, the very issues for which our stories act as a form of shorthand. Each time we reshape our narratives to accommodate these story-beats, we risk obscuring uncomfortable truths while convincing ourselves we are telling them. The result is a story that may feel satisfying to tell and compelling to hear, but that does not fully reflect the messy, non-linear nature of trauma, healing and human experience.

The culture that has grown up around the public airing of adversity also comes with built-in assumptions that aren't as safe as they appear. We're often seen as selfless crusaders for a cause and reduced to a set of knowable and noble motivations. This is not to cast doubt on the intentions of storytellers – most of us aspire to do good – but there are risks when we are essentialised according to the stories we tell. Some may never assume, given our stated desire to help others, that we may also hope to get something out of it for ourselves. This is no bad thing, but it must also be understood that where storytelling may lead to further opportunities, new incentives and risks arise. Prominence in our communities, media appearances, public-speaking gigs, and, in my case, book deals and TV shows, can tempt us to wield our stories strategically. I'd be lying if I said I wasn't aware of how my 'story', and how I tell it, can open doors. Like anyone in a professional setting, we who share so much of ourselves are not always without our own agendas. Like the rest of you, we have a need for financial security, and a desire to be well thought of in our community. We often come with a chip or two on both shoulders and some long-standing points to prove arising from old wounds. We may also harbour political objectives as campaigners which open us up to the pettier side of community politics; beneath the earnest calls for systemic change, there may be grievances to vent, scores to settle or personal resentments to air out behind veils of social justice concern. And while our testimonies deserve sensitivity, treating every experience as sacrosanct – beyond question – also has implications. As long as individual personal testimony serves as a tool for policy or civic dialogue, this narrative shaping becomes a double-edged sword. When we share, it is often forgotten that we can only really ever speak for ourselves. I have used my platform countless times to speak on behalf of various groups, from the working classes to alcoholics and addicts, and while I'm certain many who count themselves among those groups identify strongly with my personal experience and attendant opinions, I'm sure just as many do not, or wouldn't if they'd heard of me.

Despite the affirm-only ethos that pervades, this movement – which prizes personal accounts above all else – is not without its sceptics. There's a risk – and it's not a small one – when people with lived experience are positioned, or come to position themselves, as 'experts'. Not experts in our own lives – that's a given. But experts in public health. In clinical decision-making. In systems design. While we must have a voice in debates that impact us, we must also exercise humility and resist the temptation to overvalue our first-hand experience. Adversity grants us insight into how pain feels, how systems can fail, or how stigma or public ignorance on a given issue may aggravate suffering. It doesn't, on its own, grant us special rights or privileges to shape policy or clinical care. And when the lines get blurred – when personal narrative is mistaken for universal truth – there's real danger. While blindly deferring to authority is never wise, it is equally naive to assume that any professional or expert who disagrees with us on a given topic does so because they are ignorant, dispassionate or bigoted. While I don't doubt the abundance of ineptitude and prejudice among highly qualified professionals in public life, not everyone who pushes back on the impassioned claims we make can be easily dismissed as self-serving or uninformed. On the contrary, it takes great courage to challenge a prevailing lived experience narrative. Not least in the face of tremendous informal social pressure to submissively defer to sanctified personal truths, out of fear of being seen to invalidate lived realities. In such a climate, where we come to over-estimate our expertise while reflexively doubting the wisdom and intentions of professionals, we risk building services around stories, rather than evidence. We risk elevating the loudest or most relatable voices over the most rigorous ones. And perhaps most worryingly, we risk flattering survivors into roles they're not trained for and then blaming them when things go wrong. While our first-hand experiences often drive a deep passion and commitment for change, we may at times grow overzealous or strident, creating strained dynamics in public debate which actually undermine our aims. And part of our frustration will arise from the notion that our lived realties are not being heard or taken seriously, when perhaps, sometimes, they are simply being given their appropriate place, and what we are really experiencing is the reopening of old wounds of feeling unseen. The arguments of sceptics must be taken seriously as they often contain kernels of truth – even when made insensitively. Critics argue, correctly, that first-hand testimony, when taken uncritically, often smuggles unscientific, politically motivated anecdotes into serious discussion and debate. They balk at the idea that those of us with traumatic histories are automatically granted 'victim status' in public conversations, especially when our stories go unexamined, or carry political implications. This is especially so in the heat of any debate that touches the delicate matter of identity, where certain experiences are privileged over others according to political ideologies or tribal loyalties – one person's trauma is another's cynical weaponisation of adversity for political gain. Furthermore, organisations often use our stories for political leverage, relying on our suffering as a shield against criticism. For many organisations, lived experience is the latest must have accessory. By placing us front and centre in prominent campaigns, a false impression may be created that people with direct experience of adversity exert more influence within said organisations than they really do, when the truth is often that those of us sharing our experiences are rarely involved in any meaningful decision-making. Instead, we're left to trade our adversities for opportunities which may or may not materialise. For some, this represents real progress, enhancing both the lives of those who've experienced adversity as well as the organisations platforming them. However, it must be understood that much like we may tell ourselves incomplete stories about our own lives, the wider sectors facilitating those stories, whether charities or media, often tell themselves equally tall tales about why certain stories are selected while others are left on the cutting room floor.

Not all stories of adversity are created equally. A story's usefulness often depends on the political agendas it serves. Some campaigners, for example, argue correctly that the voices of ethnic minorities matter in debates about racism (which, of course, they do) but should any member of that very minority raise their voice to oppose the aims of those campaigners, they may see their ethnic minority status, experience and insight dismissed, minimised or even recast as fabricated or harmful. We see this principle play out across many social justice debates and campaigns – protected status grants special privileges as long as political lines are toed. I've experienced this myself, where my attempts to acknowledge the role of personal responsibility in navigating poverty and addiction have been described as 'hackneyed right-wing gotcha moments' by left-wing middle-class academics. The same academics who would likely cite me in their qualitative research and deploy my narrative as a political Trojan horse, if only my opinions fit themselves to their ideological storyboards. This selective storytelling is not just about individual narrators emphasising certain plot points – it's also about the organisations, campaigns or movements, amplifying those narratives to further their own goals – and ignoring, minimising or tearing down those that undermine them. Here, again, we see the potential for the lines between truth and performance to become blurred. But there's another, deeper issue here, so subtle that few ever consider it: what if our trauma itself predisposes us to overshare? What if our need for safety, security, and validation makes us unwitting participants in our own exploitation? What if our chronic impulsivity and lack of healthy boundaries prime us for public exhibitionism? By placing the burden of visibility on people who may be vulnerable because it's politically expedient, organisations and campaigns risk perpetuating harm contrary to their intentions. Their affirm-only ethos is convenient as it relies solely on a survivor's willingness to participate when perhaps we require protocols to truly ascertain any potential risks posed by public storytelling before it happens – and chains of accountability should those safeguards fail. Once we share our traumas in the public square, they are no longer ours. Our stories become public property, reshaped and consumed by others. Most stories won't receive mainstream public prominence, but any level of visibility at all may prove too much for someone still living with the effects of trauma. The desire to help, to gain love, affection and security, can overshadow doubts about the risks of exposure. Quite frankly, most people living with active trauma would never dream of stepping into the spotlight unless compelled by a wound so deep they couldn't stop themselves. This brings us to a key paradox: visible advocates are often seen as representatives of entire groups when, in reality, we may be a distinct subset – not voiceless masses, but individuals often driven by a need for validation, recognition or a sense of agency. We want to be seen and heard in ways most trauma survivors (or people for that matter) don't. This creates a broader issue in the industrialisation of personal storytelling: stories told by charismatic or compelling narrators, which fit well-worn frameworks and formulaic structures, rise to prominence, but often at the cost of authenticity. The danger here is that audiences and policymakers may take these highly revised personal stories as representative of everyone with similar struggles, distorting public opinion and policy over time. While first-hand testimony is valuable, it cannot replace more systematic approaches, such as large-scale surveys or close work within social settings. We'll explore in greater depth the costs of public disclosure later, but its current dominance stems not from its undeniable usefulness but from its accessibility – it's relatively cheap and simple to produce, there's no shortage of trauma, and many people believe sharing their story is inherently a good idea.

Later in our journey, we will return to many of the issues outlined here, in greater depth. What's really worth paying attention to, however, isn't always the polished narrative of overcoming adversity, it's the risks we take when we mistake the fleeting catharsis of being seen or feeling heard for the hard work of healing. It's the loss of truth in our lives as we reshape our hero journeys for public consumption, and the civic and media sectors, indulging in these narratives frequently, that conflate individual experiences with universal insights. Granted, quite often it comes off well and a world of good is done, but sometimes it doesn't – these are the stories you are less likely to hear. The drive to tell a captivating story may overshadow every other concern. Sharing our experiences is natural and important, but it must be done with care. There's a time and a place to do it safely. Once that genie is out of the bottle, there's no putting it back. Trust me. As someone with lived experience of being 'lived experienced', I can tell you: the greatest risk isn't that your voice won't be heard. It's that it attracts so much attention, your life becomes a performance – one that confines you to the very traumas you hoped to escape.

'Vulnerability without boundaries is
not vulnerability. It can be desperation,
or even trauma reenactment.'

Brené Brown, *Daring Greatly* (2012)

CHAPTER 4

The Debasement of Pain

*What are the risks when anyone can
claim they have trauma?*

While I dedicate much of this book to bemoaning the burdens of being a public face of poverty, addiction and trauma, the truth is, I don't know who or where I'd be without this story. I don't know how my life would feel or how it would have turned out without these wounds framing and defining it. And I'll even let you in on another little secret. One I'm a little scared to admit given how often I'm portrayed as someone who has successfully overcome their difficulties: when it comes to trauma, truly healed people unnerve me. Their sincerity is cringe-inducing, their lack of ulterior motive unsettling, and their uncanny ability to maintain their intense, invasive brand of eye contact – like flowers tracking the sun's rays – borders on abuse. I resent their quiet faith in life's uncertain and uncomfortably long embrace almost as much as their indifference to the approval of others. I dislike how their presence magnifies my own shortcomings. Give me a cup of warm coffee and a room full of drug-addled, neurospicy fuck-ups and I'll be happy as a pig in shit. In those rooms, I'm the Noam

Chomsky of self-help. But sit me down across the table from someone who actually knows how you get well from this stuff and watch me squirm in my seat. I deal with a form of imposter syndrome many in the broadcast-your-pain community know all too well: I know I'm not quite the person I want everyone to think I am. Why I keep seeking out these paragons of healing who elicit feelings of spiritual inferiority in me remains a mystery. Perhaps it's just a cleaner form of self-harm, or maybe some humbler, wiser part of me knows that despite my award-winning performance as a recovered alcoholic addict turned success story, I'm nowhere near as 'recovered' as I'd like people to believe. In a sign of the times, someone as quietly anguished and insecure as me has somehow become, to many, a poster child for how a life should be lived. Still, despite my occasional discomfort around those I assume are further along the road of healing than me, I actively seek their company, convinced they must know something I don't.

*

Today is one of those days. I'm preparing for the arrival of a dear friend and fellow former fuck-up in my usual fashion: chain-smoking cigarettes, chugging energy drinks and scrolling social media for signs that my existence matters to someone. Anyone. I'm in the green room at Stand 3, my venue for an eight-day run at the Edinburgh Fringe. My old friend James is already here, making himself at home. Predictably, James is exactly where he said he'd be, at precisely the time he said he'd be there. Like all truly healed people, he's dependable – never dishevelled, never flaky, always even-keeled. James has been a cornerstone of my sobriety since 2015, stepping in to support me after my first attempt at getting clean ended in relapse. James was my sponsor – a mentor of sorts who provides guidance on the road to recovery from addiction – and I'm just one of hundreds he's helped. But unlike the kind of 'help' common in online trauma discourse, James doesn't just affirm my every half-baked assumption. He loves me enough to risk offending me by telling me the truth as he sees it. Over time, we grew close enough that after my 2022 relapse (the one I didn't announce publicly) we agreed it was best for me to find a new sponsor. I needed a new experience, and James had begun to be impacted by my relapses, not least because I was now a public figure whose slip-ups often found their way online. Despite stepping back from the sponsor role, James and I remain close, and his presence often has an anchoring effect – useful on a hectic day like today. As we chat in the cramped dressing room, another guest arrives: Suzanne Zeedyk, a child psychologist and campaigner who has profoundly influenced Scotland's ongoing conversation about trauma. Suzanne's gaze is unnerving – when she looks at you, it feels as though she truly sees you. When her eyes lock on, a piercing searchlight is beamed into a dark corner of your soul. Naturally, this makes me want to avoid her gaze entirely. Who wants to be seen by someone who possesses genuine insight? I'd much rather be admired by people who, like me, aren't as well put together as they believe, and as a result of their lack of insight, may mistake me for a veritable spiritual giant. Taking a show to the Edinburgh Fringe is a dream for many performers, and like most misguided ideals, the sobering reality is worth complaining about, publicly. Edinburgh itself is a city of contradictions – part twenty-first-century tourist trap, part eighteenth-century financial hub and part living museum of the disease-ridden dark ages. For every grand monument, there are dozens of boarded-up shops and for every rough-sleeper, countless vacant Airbnbs. It's a factory reproducing trauma on an industrial scale where many have become so inured to the sight of the wounded, they no longer bat an eyelid. August brings a special kind of chaos to the city, with Edinburgh transforming into a breeding ground for financial extortion. Performers, me included, like to tell ourselves we're here for the love of our craft, but really, we're here for validation. We're here because we need to know we matter. This year, my show – *Trauma Industrial Complex* – is sold out, thanks to the public's insatiable appetite for all things trauma. In recent years, 'trauma storytelling' has become a genre unto itself and tens of thousands will walk past dozens of homeless people and desperate substance misusers without a second thought – on their way to watch comedy, music and drama (prefaced by trigger warnings) delving into the need for greater compassion and sensitivity for the still suffering. With the working classes and the poor – statistically likelier to bear the wounds of trauma – largely priced out of therapy, they look on at the Trauma Industrial Complex from the outside, leaving plenty of surplus oxygen for the urge-surfing affluent, shopping for a hard-luck story, to suck out of the room – making this circus the perfect companion for the Edinburgh Fringe. Hundreds of shows this year deal with trauma in a variety of disciplines and tones. None of them, however, are approaching the nebulous topic quite like mine. My goal is to gently critique the sacred cows of storytelling and encourage reflection on the implications of trauma's commodification. I want those invested in trauma discourse to consider whether the stories they facilitate or tell might be incomplete – and whether editing out the 'bloopers' could be prolonging their pain or painting an

untrue picture. This conversation isn't new, but it's often relegated to whispers behind closed doors. Perhaps now it's time
to bring it into the open.

*

'Trauma is how life feels,' Suzanne begins, addressing the 180-
strong crowd, 'and then what that does to your body – where
that lands in your biology.' The audience now captivated,
she continues: 'We tell ourselves the stories of what happened, but without understanding the biology, we don't really
understand what trauma is.' Suzanne and James are prominent voices in the Adverse Childhood Experiences (ACEs)
movement, which originated from a landmark study by the
US Centers for Disease Control and Prevention (CDC) and
Kaiser Permanente in the 1990s. This study categorised adversity into abuse, neglect and household dysfunction, finding
strong correlations between the number of ACEs (traumatic
events) and negative health outcomes, such as chronic disease,
mental illness and substance abuse. While hugely influential in
normalising the lexicon of trauma, the ACEs framework has
faced criticism for its limitations, including its homogeneous
study population (participants were mostly middle class), failure to consider cultural and socioeconomic differences, and its
equal weighting of diverse adverse experiences – divorce and
physical violence both equal an ACE. Then there's the scoring system itself which, while useful in discussion among peers
with respect to the level of adversity experienced, may also
encourage survivors to draw inappropriate conclusions as to
their level of trauma, its effects and other long-term implications. Scoring creates a culture where trauma can be quantified,

which is misleading to say the least. Nevertheless, the premise – that unmet childhood needs can manifest as poor outcomes in adulthood – resonates deeply, which is why many are drawn to the framework. My own experiences align closely with this narrative. My mother, herself a victim of childhood abuse, had me at 18 – a high ACEs score in children is a predictor of teenage pregnancy. I endured chronic emotional neglect, and several incidents of physical abuse and threatening behaviour – again predicted by a high ACEs score. My parents separated, and my mother succumbed to alcoholism. By 17, I was homeless, engaged in a toxic relationship, and later hospitalised multiple times for overdoses and suicide attempts. While anecdotal, my story echoes the conclusions of the ACEs study – that early years adversity is a prelude for poor outcomes later in life – highlighting the long shadow of unmet childhood needs. James expands on how trauma manifests in the attitudes and behaviour of many of the working-class young men he mentors: 'It shows up in reactions to life – how they maintain relationships, set boundaries or cope with addiction. Addiction, in particular, points to an unhealthy relationship with oneself.' James, actively engaged in the business of helping others through community initiatives and mutual aid groups, is a proud ACEs advocate. Of course, the manner in which he's turned his life around could be seen by contrarian arseholes like me as an argument against the ACEs framework. What the ACEs study does not adequately account for is how children who endure even the most chronic or extreme neglect or abuse are not necessarily consigned to miserable futures. The biggest unintended consequence of the study, you could argue, is that it's fatalistic, and that wrongly interpreted it risks framing early trauma as an inescapable determinant of future suffering. By assigning a score to childhood adversity and correlating it with long-term health, psychological and social outcomes, the impression may be created that a high ACEs score means a predetermined path of dysfunction. This deterministic reading may overlook resilience, social support and personal agency, reducing complex human development to a set of risk factors and red flags while under-emphasising the role of protective factors – such as stable relationships, community support and personal attributes or coping mechanisms – that can offset or integrate the effects of early trauma.

Additionally, the ACEs model, when applied rigidly, risks medicalising or pathologising individuals based on their past rather than focusing on their capacity for growth and healing. It may also shift attention from structural change (addressing poverty, systemic inequality and access to care) to individual pathology, reinforcing a narrative where trauma is a lifelong sentence rather than a challenge that can be worked through.

Luckily, Suzanne and James avoid these pitfalls. They take critiques of the ACEs framework seriously and often engaging with them head-on, with open minds and humility. Their work focuses on the role systemic issues like poverty play in creating the social conditions for trauma. From early years to education and criminal justice, they understand that adversity rarely occurs in a vacuum and believe, correctly, that unmet need caused by the strain of poverty drives much of the trauma we see in Western societies. Trauma which later finds expression as family breakdown, low educational attainment, police, social work and criminal justice involvement, addiction, homelessness and, in many cases, death. Despite the long-standing relationships between the three of us, differences of opinion on the rise of trauma as a concept are emerging in our discussion, though I'm thankful to say nobody has yet been triggered. Suzanne sees the popularisation of trauma as a net positive and is proud to be part of the ACEs movement that did so much of the heavy lifting before Instagram and TikTok took over. 'I'd rather we talk about it, even if imperfectly, than ignore it,' she says. James echoes this sentiment, noting that the ACEs framework provides a safe way to discuss adversity without oversharing. 'When the ACEs movement landed,' he tells me, 'the reason I ran with the framework is because it allows people to speak safely about what might have happened.' James refers to the impulse to share our experiences with others – an essential part of any healing process from trauma. The ACEs framework, he explains, gives those who have suffered adversity a means of articulating their experiences which also protects their privacy. James found the ACEs framework helpful in his role as a mentor and campaigner, for the same reason. 'It allowed me to talk about the pain in my own life,' he says, 'for example, I could say things like, "I've had a number of ACEs," without giving people the details.' He continues: 'Not everybody has won the honour of hearing the intimate details of my life experience.' James understands the need and desire to disclose our adversities must be balanced against the risk of saying too much. Most in the trauma field who work with vulnerable people know this too.

But online, this balance is often lost. People disclose intimate details publicly, sometimes during active crises, encouraged by a culture that values gritty authenticity. This dynamic extends to media and organisations, which may sometimes exploit lived experiences without fully considering the potential harms. For example, Netflix's *Baby Reindeer* showcases what has been described as a deeply personal story about several extremely traumatic experiences but raises questions about the ethics of publicising such narratives and how to ensure there are adequate safeguards – it became a media circus that arguably placed vulnerable people at risk. People with trauma may attach too readily, due to unhealed wounds, and this creates a risk of exploitation, where the catharsis of sharing may be prioritised over the storyteller's wellbeing. Of course, if someone believes they are well and conveys to those who might facilitate their story that this is the case, then the risk of exploitation comes not from a third-party but from the storyteller themselves – a person who may lack necessary insight into both their own story and their fitness to disclose it.

The greater risk, however, is that those of us who share our experiences of trauma publicly in the mental health space come to be regarded as the sole authorities on the subject thanks to the special status granted us in trauma culture – this I believe is dangerous. I put to Suzanne and James my view that while those of us with experience of trauma have something of value to contribute, this special status may actually constrain the utility of our stories. In a culture where so much value is placed on highly subjective interpretations of trauma, and where clear incentives exist to storify adversity on the assumption that public disclosure brings catharsis and healing, is a conversation powered solely by stories which are rarely scrutinised really about recovering and inspiring change or may it also be about aesthetics? 'I see your question as essentially pointing to a wider question,' Suzanne says: 'Which kinds of risk do we want to take?' She continues: 'It is true that, once upon a time, the word "trauma" wasn't in the public consciousness. "Trauma" would have been a specialist matter – a matter for "experts". Nowadays, the word, concept, process is everywhere, and anyone is deemed to be able to speak on it. Lived experience of trauma makes you an expert. It gives you a voice.' Suzanne acknowledges that we find ourselves in a transitional period where, in her words, 'the basis of legitimacy has shifted', but that subjective knowledge can be as valid as objective knowledge. Had this claim equating objective knowledge with personal interpretation been made by a sociology student on the campus of a Russell Group university, I'd have quietly dismissed it out of hand, but Suzanne is a qualified scientist. She takes my concern that trauma may be rendered meaningless in the current discourse and runs with it, almost as if to demonstrate that she empathises with my concern, while teeing up her own counterpoint. 'If everything can count as trauma,' she says, 'then what is "not-trauma"? Trauma was originally meant to signify something more extreme, something non-normative. If the extreme has become normalised, then the concept of trauma ceases to have meaning. The concept of "trauma" sort of dies. That is a fascinating possibility. Democratising "trauma" killed it.' She then flips this hypothetical on its head entirely and poses an alternative scenario, drawing on historical examples of trauma being minimised by dismissing or discounting lived realities: 'The alternative risk is that traumatic experiences are overlooked. Actual trauma is regarded as not-trauma. This is the risk of denial. Our society does this all the time. The history of trying to improve children's lives reveals that. In the 1950s, when campaigners tried to stop children's hospitals from being [what we would now call] "emotionally traumatic", the medical system pushed back and said that the care they were offering was fine. In the early 1900s, when labour campaigners tried to change factory conditions from being [what we would now call] "traumatic",

employers said it was fine. In the 1960s, when education campaigners tried to change corporal punishment to stop [what
we would now call] "traumatic practice", teachers said it was
fine. All of these people [adults] in power insisted that what
was traumatic was not-trauma. They insisted that the suffering the children were experiencing was fine, was normal,
was acceptable, was OK, was not-trauma. And so, the adults'
accounts about what counts as trauma (like you say of subjective stories) were not reliable either. Because the children
were suffering. And the adults said they were not. And the
suffering (the trauma) of childhood grew into consequences in
adulthood – which became mental and physical health problems in adulthood that no one understood or took seriously.
The mental health consequences of being regularly strapped
at school? Of having been prevented from seeing your parents when you were in hospital? Of having been forced up to
work 14-hour days? All of these actually traumatic experiences
were denied.'

It's a testament to Suzanne's confidence in her beliefs that
she is able to show such generosity of spirit in the face of my
argument – that everyone talking about trauma is dangerous
– which, essentially, strikes at the very core of her philosophy.
I begin to wonder if perhaps I too have been captured by the
Trauma Industrial Complex, but rather than wrongly misdi-
agnosing myself, I've grown overly cynical about the harms it
poses and less open to the potential benefits of trauma being,
as Suzanne reframes it, 'democratised'.

I begin to wonder what James thinks of my stance, which I'm
sure he has some reservations about given his role as Scotland's
leading advocate for trauma-informed approaches in criminal
justice.

'History is littered with examples of belittling, denying and ignoring human suffering,' he says. 'From the literature on shell shock after the First World War, the social distrust of army veterans who did not present as "normal" was common. To the ACEs data, derided by many as the study population was made up of white middle-class Americans. To the recent flippant statement made to me, "one man's trauma is another man's Tuesday."'

James, like Suzanne, is keen to highlight what they both believe is a dangerous historic impulse – the societal instinct to dismiss, diminish and disregard trauma, often under the guise of scepticism, intellectual superiority or sheer discomfort. He tells me that talking about ACEs science and research was one of the most difficult things he has done.

'The kickback came from the very institutions and people who I believed would be my persistent and able advocates – academia and lived experience.'

This wasn't a predictable kind of backlash – it didn't come from the usual critics, the hardened sceptics who dismiss trauma research outright. It came from inside the very spaces James assumed would get behind it. Perhaps this is an example of what happens when trauma gets politicised, gatekept and fought over like territory. James wasn't some outsider making wild claims – he was simply amplifying research that showed what many already knew: that early trauma is not just something you 'get over'. It leaves traces, patterns and long shadows that shape health, relationships, behaviour – even life expectancy. And yet rather than being welcomed as another piece of the puzzle, the ACEs framework became a point of contention, a battleground. 'It never came as a surprise,' James admits. And that, to me, is perhaps the most damning part

– the fact that pushback against trauma discourse is so pre-dictable that those who speak up are already bracing for it. Because it's one thing to be challenged by those who simply don't believe trauma is real; it's another when the resistance comes from those who claim to be on the same side. But James doesn't let it deter him. In fact, he sees it as fuel. 'It aided my agenda – to get people to lean in and amplify the awareness of the trauma in our culture and how we as a society need to get better at prevention and the recovery provisions that treat it holistically.'

The refusal to take trauma seriously is, in his view, nothing new. The same shell-shocked soldiers now cited by sceptics like me as evidence of 'real' trauma were, in their day, cast aside as weak or defective. Indeed, it was only due to the efforts of health professionals and advocates that PTSD was eventually recognised as a clinical condition. Even today, the overuse of trauma-related terminology does not come from nowhere; it is, at least in part, a rebellion against the long-standing ten-dency to deny suffering altogether. And yet where does that leave us? If trauma is stretched so thin as to cover every incon-venience, does it not risk losing its power to name true harm? When everything is trauma, nothing is trauma. This is the core concern – that, in an age of over-identification and mass self-diagnosis, the weight of genuine suffering is diluted, trivialised and repackaged into something marketable.

The phrase 'one man's trauma is another man's Tuesday' may be crude, but it captures a widely felt unease – the creeping sus-picion that trauma has become a linguistic free-for-all, stripped of its historical, medical and ethical gravity. James, however, pushes this discomfort further. 'What if we flipped that narra-tive and framed it like this: "one child's rape is another child's

Tuesday"? Would it be an acceptable statement in any setting to view the severity of traumatic experiences through this lens?'

The rhetorical force of this comparison is undeniable. It exposes a brutal contradiction: if we are too quick to label everything as trauma, we risk undermining its significance – but if we are too cautious, too sceptical, too insistent on policing definitions, we edge dangerously close to complicity in silence. Both James and Suzanne make a persuasive argument that the greater danger we face is repeating the mistakes of the past where real pain is swept under the rug by gatekeepers. James then addresses my concern about survivors taking ever more public platforms so early in their recovery journeys.

'My disclaimer has always been that if just talking about ACEs and trauma was the solution, I'd be a paragon of healing. But the other side of that is, if we had to wait until people were healed before we gave them voice, we would all still be waiting. For most people who experience trauma, it is not a single incident – it is an environment, so it cannot be democratised, because it has never been distributed evenly. I think we risk analysis paralysis if we get stuck on who should be the key stakeholders in trauma discourse. After all, humanity is built on stories of overcoming adversity. It is more dangerous not to talk about it.'

If trauma is, at its core, about injustice, James contends, then speaking about it – however imperfectly, however clumsily – is part of the necessary process of truth-telling. This is not a flawless movement; the Trauma Industrial Complex has unintended consequences, commercialised distortions and real dangers of excess. But the alternative – the old tradition of denial, suppression and shame – is surely worse. James and Suzanne make a powerful case that people who disclose real trauma – abuse, neglect, violence – often risk alienation, not validation – an important point I begin to fear I have overlooked in my agitation at trauma's seeming commodification. The consequences of speaking up can be exile, fractured families and institutional retaliation – all of which I have experienced myself. Suppression, as James notes, is not a modern invention – it has been a survival mechanism for generations. He proceeds with another crucial distinction: trauma is not simply the presence of suffering, but the absence of recognition. 'It must be recognised that a survivor telling their story is also about the struggle for the recognition of an experience of injustice,' he says. 'That is what trauma is – an injustice – and for many, that process is cathartic.' James, much to my relief, does share some of my anxieties about trauma being misapplied, and actually cites the Western media framing of Covid and lockdown as 'traumatic' as a prime example. 'The pandemic was touted as a collective traumatic experience for society by many,' he says, 'but I do not believe it was.' James continues: 'Firstly, because there is no evidence to back such a claim, and secondly because everyone was able to talk about it. It was not a taboo subject such as the ACEs study or sexual abuse.'

James, like Suzanne, acknowledges the concerns I express; the risks of over-identification, of diluting trauma's meaning, and the spectacle of public storytelling are real. 'People with lived experience need to be very cautious about public exposure in storytelling and their motives for doing so,' he warns. 'There is a world of difference in sharing from a place of unresolved trauma and from a place of post-traumatic growth.' Obviously, I'm glad we're finding common ground. After all, this distinction matters and lies at the heart of my own deep dive into this topic. 'Organisations have an ethical duty to assess and formulate safeguarding plans that have the survivor's best interests at heart,' James insists. 'Autonomy must always belong to the storyteller.'

As our event draws to a close, the hour flying by as it always does, I'm reminded that despite playing for laughs occasionally by mocking the urge-surfing trauma-curious tendency to misapply labels and frameworks, I'm less concerned about the prospect of people believing they have trauma when they don't – and far more about the risk to those drawn to public exhibitionism who don't yet grasp how traumatised they really are. Telling your story before you're ready, before you've processed its meaning, before you have any distance from it, can be dangerous. There is real potential to re-traumatise. It may render you vulnerable in ways you don't anticipate. And when organisations parade people's lived realities for their own agendas – without proper safeguarding, without true consideration for the survivor's wellbeing – then it stops being about justice or healing. Like many of the shows running at this year's Fringe, trauma instead becomes a spectacle, where pain is performed rather than truly processed and released. And in this parade – not much ethically better than an exploitative daytime talk show – countless others are led by bad examples, to the precipice of their own public disclosures without any real idea of what new adversities may await them as a result. Perhaps outsourcing the safeguarding to the very Trauma Industrial Complex that benefits endlessly from our lack of self-awareness or boundaries is only part of the answer. It seems to me that the best defence against making ourselves vulnerable to exploitation, by others or ourselves, is to take responsibility as storytellers for how we move in this space. Any solution that rests on assuming the good intentions of others, no matter how well meaning, is only half a solution. We must become willing to take full control of our stories and complete responsibility for the implications of telling them.

'Trauma has no hierarchy, no tidy lessons, no linear arc. But the world wants it packaged. Palatable. Resolved.'

Carmen Maria Machado,
In the Dream House (2019)

CHAPTER 5

Full Disclosure

How do we take control of our stories?

I know a thing or two about trauma. As unpredictable as it is, if trauma does one thing reliably, it's humble you. That's my experience. Recovering from trauma involves a lot of acceptance that feels extremely unfair. The most obvious injustice is that you likely did nothing to deserve the wound that was inflicted on you. One day, life is normal, boring, predictable, and the next, a traumatic event is dumped on your emotional dining table like fresh roadkill. Whatever plans you had – forget it. You're on trauma's time horizon now, and for every day that seems to stretch a week in length, there are twice as many that fly by so fast you could swear you dreamed them. The simple tasks you feel you can no longer perform, the commitments that go unfulfilled, these only add to constant self-loathing. But it's the acute sense you are as good as useless, appearing weak of will and feeble of mind to anyone unfortunate enough to catch a glimpse of you when you're 'activated', that really begins to grate. Trauma is indeed humbling, though humility isn't my strong suit as you can probably tell. So even when that sleeping giant within me was rudely awakened recently, I just tried to power through – with mixed results. At the inaugural event of my Trauma Industrial Complex campaign in 2024, a few months before the Fringe, I asked my guest, Miriam Taylor – a world-renowned psychotherapist with her own experience of trauma – to define the elusive term. She obliged with this: 'Trauma is the imprint of an event, situation or relationship that threatens our survival and overwhelms our capacity to cope on our own.' It was a definition I recognised intimately, not least because I had arrived at the event with Miriam fresh from a therapy session in Glasgow, exhausted and visibly shaken. When I began writing this book in the winter of 2023, I optimistically aimed to deliver the first draft by the following March. But two months before the deadline, my progress was derailed. Much of that time exists now as a blur, though I vividly recall the bitter irony of finding myself, mid-project, receiving a visceral and humbling reminder of what trauma really feels like. I can confirm it's no fun. There's the traumatic event itself, obviously – the moment terror overwhelms you – but what really breaks you is the way its impact ricochets through your nervous system, long after. When you live with active trauma, the flood can overwhelm you without warning. Even as I write this, the words pour from an open wound. Much of the material for this book – as it stood before this calamity – had to be thrown out or reworked. It no longer made sense as it had been written from a perspective where trauma was an abstract idea: an experience I recalled vividly but was not entirely possessed by as I am now. One day, that all changed, and trauma was no longer a memory I merely called back to, but a bodily condition that dictated my existence. The specifics of what happened aren't relevant here – perhaps I'll overshare them later – but trust me when I say I know what trauma feels like. For me, the rupture begins in my abdomen, an overwhelming sense that something is deeply wrong. The sickening sense that I am not safe. That feeling arises without any discernible cause, reminding me that trauma is less a story about what happened and more a physical experience. It's the shock of being disabused of the illusion of safety. Once inflicted, that wound demands awareness and understanding to heal – processes which are repeatedly derailed by the cascade of physical and emotional chaos that follows. Triggers and emotional floods wreak havoc on your body, mind, and relationships. Psychologically at sea, you reach desperately for whatever remains of the shipwreck you can grasp just to keep your head above water long enough to catch a breath. It helps when you aren't alone, but often the worst episodes unfold when no one else is around to offer a word of comfort. Those moments, of utter isolation as this entity takes you hostage, are among the loneliest you'll ever experience in life. Popular narratives about trauma recovery tend to present it as a three-act arc, something linear and resolvable. But that's not my experience. Trauma is more than a painful story; it's a physical onslaught. At the edge of sleep, a sudden shot of adrenaline often jolts me awake, leaving me alert for hours. I get out of bed and pace the house, searching for clues as to where the next threat might be lurking. Some days, I forget to eat, relying on quick fixes like chocolate or cereal to maintain basic functions because the thought of preparing, tasting, chewing and digesting a proper meal feels overwhelming. Sleep deprivation then amplifies the emotional and psychological chaos of anxiety. Every day, I face a barrage of triggers – some predictable, others catching me completely off guard. Sounds, smells, pieces of clothing, music, even a date on the calendar can set me off.

One day, I'm coping. The next, every joy feels like a curse, and every responsibility like a mountain I'll never scale. I lost 4kg in a couple of weeks – something I couldn't achieve on any fitness plan. I was hospitalised for an oesophageal spasm, which feels alarmingly like a heart attack. I've endured nights without sleep, days spent in dark rooms and unusual physical pain: nerve pain, headaches, backaches, joint pain. Things have improved recently thanks to those around me learning how best to support this particular type of trauma, but I've learned there are no guarantees. Some of this work I must do on my own, even if others kindly and patiently walk alongside me.

What triggered this spiral wasn't a dramatic event like a car crash or a death. Bombs were not raining down on me from the sky. It was an innocuous detail revealed in an everyday conversation, a detail that sent shockwaves through my body before my mind could process its implications. Before I had even begun to fashion a story of what happened (a mental model to explain the sheer ferocity of the physical symptoms) I was upstairs, shaking violently behind a bathroom door, my body trembling until my bowels gave way. Strangely, this traumatic event was preceded by weeks of increasingly poorer health. My body seemed to anticipate the crisis well in advance, as I'd become gradually more unwell. These were subtle signals my mind refused to acknowledge; my sense that some important aspect of my life was not in order was suppressed and written off as the rumination of a caffeinated mind. Perhaps that's just another story I'm telling myself, but eventually, the physical signals became unignorable. Trauma would have its day, and I would finally be humbled. One evening, my body sounded it's unequivocal two-minute warning, throwing me into full-blown panic, thus setting in motion a chain of painful events from which I'm still recovering. The only grace I was shown by life in that moment was the fact I'd been here before. These symptoms weren't new; they're ghosts from my past, roused after years of dormancy. I thought I'd exorcised them, but trauma's reality is often cyclical, not linear. You never really 'recover' in the way terms like 'healing' suggest. At best, you learn to accept the limitations trauma places on you and move around within it when it's activated. The episode forced me to reconsider much of what I'd been telling myself about trauma – and the discussion and debate surrounding it. When I started this book, I was fuelled by irritation at the mental health minutiae dominating social media. Typically, I was knocked square off my high-horse soon after by a pride-levelling calamity – life's way of reminding me I don't know as much as I often think. On the day I met Miriam in Edinburgh, I had just emerged from an emotional flood – an overwhelming surge of painful thoughts and intense feelings – and her presence offered a stark reminder of the role others, with the skill to emotionally attune, can play in soothing the pain. Miriam's presence was steady and assured. She could see and feel the trauma radiating off me. Strangely, simply feeling seen and felt can be enough to pull you back out of the nightmare for a while. Oh, there I go again, trying to sound like an expert. What the fuck do I know? The day I met Miriam for the first time, I had less than an hour to pull myself together as we prepared to stand before hundreds of people … to talk about trauma.

*

Now let me say it again, for the fuck-ups in the back: trauma is not a story about what happened; it's the wound felt by the nervous system. This is an important distinction that much of our current conversation about trauma, with its focus on narratives, often fails to account for. That said, despite the problems and risks associated with this culture, people will continue to tell their stories, and, in my view, we spend energy foolishly resisting this reality. Who am I or anyone else to pontificate as to how others should or shouldn't vomit their private lives up all over anyone who will listen – I'm not exactly a black-belt in keeping my own powder dry, am I? Where there exists demand for personal anecdotes of trauma, it will be supplied. The questions we now face are about how useful incomplete stories are beyond fleeting catharsis and entertainment, and how to tell important stories safely when they are shared publicly. I am now of the mind that the best defence against blowback from storytelling is for storytellers ourselves to own our shit – an area we'll explore in greater depth in Part 3. We must enter this culture with true awareness and intent rather than being chaperoned, mollycoddled and pacified like docile circus animals. In my view, preparation begins by asking ourselves why a story is being told in the first place. That may seem a silly question, but is it? Take my first-hand account of trauma in the previous section for example: what is its purpose? Perhaps it's my attempt to authenticate myself in your eyes, as someone who understands trauma. Or perhaps it was deployed strategically to draw you further into a story, much like a screenwriter uses rising tension to further immerse the audience in their narrative. Maybe it's neither or a bit of both – that's not your concern. As the storyteller here, I am fully aware of why this story is being told, why it's structured as it is, what I hope to achieve in telling it and what the potential consequences might be. Every line in this book has been checked and double-checked. Every revelation considered carefully. I have robust personal and professional support structures around me that most people don't. Whatever fallout may come from anything printed in this volume, you can be sure it'll be handled and that I'll be OK. But can we say with any certainty that the highly personal narratives currently spooling across the Trauma Industrial Complex arise from a similar intentionality, steadied by the same guardrails? I doubt it.

When we tell stories about our lives, generally speaking, we either do so with conscious awareness or we share in a form of narrative autopilot. These spontaneous, improvised, wholly imperfect accounts of ourselves and our lives are natural and carry few consequences. Most people can get away with skimping on the details because most stories quickly become buried under new narratives and most people, truth be told, are too busy telling stories about themselves to retain the details of yours. Everyone runs with a story about their life and not all of that story is true – that much we know. Where stories like mine differ is that those stories aren't as quickly forgotten. Go and ask the drug addicts, or the victims of knife crime, or the morbidly obese who at some point agreed to tell their stories in the papers or on the news, how they feel now, years after being portrayed unfavourably in public. Many personal narratives about trauma are told with great intentions arising from magical thinking. Too many storytellers lack a firm enough understanding of the cultural terrain, or the transactional reality underpinning the economics of storytelling. There is little real awareness on the part of storytellers about why they are even telling their story, or what the consequences of it may be. This book, while touching on some important wider issues briefly, is primarily concerned with addressing one specific conundrum: how do we protect the integrity of lived experience stories and the safety of storytellers?

I've been telling myself a story about trauma recently, too. The story about how the conversation around it is wildly out of control. The story about how the term is endlessly misapplied. Trauma has been a controlling idea of my story for as long as I can remember. Yet, oddly, when I began writing this book, I couldn't recall having ever being officially treated for it. Imagine that: I had lived most of my life believing a narrative about trauma and its impact, even though its presence had never been, to my mind, formally verified. That might seem an inconsequential detail but playing fast and loose with the basic facts of our narratives around trauma is more common than many of us realise – or are willing to admit. I eventually sought to confirm whether my story was true – as we'll explore in the next chapter – and then came my own minor calamity not long later. For now, however, I want to impress upon you, whether you're a storyteller, a facilitator, a consumer or even one of those 'normal' people outside the Trauma Industrial Complex, that the best protection against blowback from an incomplete story is willingness to perform a page-one rewrite. I'm not suggesting the traumatic event didn't happen – it did, and you have a right to talk about it however you see fit. But going public, at any level, requires a little more diligence. The human mind is not famed for its commitment to accuracy and sometimes we subconsciously fill in the blanks, attribute aggravating factors like intent or culpability to others, or draw wide-of-mark conclusions as to what it all means. All people do this, but not everyone subjects their limited interpretation of key life events to public scrutiny. That's why we have to be careful.

Throughout our lives, we run with various stories about ourselves and our circumstances – not always because they're true, but because they serve us in some way. Take the penis-in-the-drawer story. Everyone has a tale like that: a yarn spun to deflect embarrassment. Then there are the white lies of omission, threads woven to create a façade we hope will influence how others see us. For instance, when Gabor Maté suggested I may have ADD after meeting me backstage at a London literary event, I immediately wove it into my narrative. How awesome is that? The king of trauma discourse himself offered me a possible neurodivergent label just as neurodivergence approached the very apex of its trendiness. By the laws of association, being described as potentially ADD by Gabor made me, for a brief moment in 2019, the coolest childhood-trauma victim on the face of the earth. Did I ever get tested? Fuck no. Did he know me well enough to apply such a frame to my behaviour? I don't know. In any event, it suited me when he suggested it. The idea I had ADD felt true, so why not just run with it? That story appealed to my ego far more than the real story which no one knew at the time: I was dabbling with over-the-counter painkillers to relieve the immense stress of my story so suddenly becoming public property. The buffet offered in the Trauma Industrial Complex can be a bit like that. Labels, frameworks and diagnoses function too often like novelty products or limited-edition mashups of our favourite brands. We feel drawn to them but couldn't say with any certainty why. That justification usually comes later, and not always for the best reasons. These self-deceptions, omissions and leaps of faith are rarely malicious. But over time, they may accumulate. They pile up like cardboard boxes in a hoarder's loft, cluttering our sense of self. Like a guitar ever so slightly out of tune, only the keenest of ears may detect something's amiss but deep down, we often know the cause of the disharmony – we aren't living in the truth. When the story we tell ourselves about trauma is incomplete, and we are being bombarded by contradictory and sometimes even harmful information about our own pain, we risk making matters much worse. Some of us, me included, can become so untethered from reality that we pay a heavy mental health cost, inadvertently causing distress and even trauma in those around us. In today's trauma-affirming culture, questioning someone's narrative may seem cruel. But if our identity is tightly interwoven with our stories, shouldn't we interrogate them? Kevin Hines thought no one cared when, in truth, everyone did. Leeann White insisted she 'can't do anything anymore', but was that entirely true, or was that just how she felt at that stage of her non-linear recovery? And as you're about to find out in Part 2, important elements of my own story – the one I've been running with most of my life – involve similar unconscious deviations from the facts. Even if you've never experienced adversity such as that described in this book, can you honestly say every story you tell about yourself and your life holds up to scrutiny? I doubt it.

In Part 1, we explored the rise of trauma culture and the risks posed by its industrialisation. We examined the eruption of public interest in trauma and the powerful notion that telling your story is essential for healing. We established the risks posed by the rise in catharsis culture and the industrialisation of lived realities. We have also heard the important counterpoint from Suzanne that the alternative – trauma becoming once again ringfenced by a specialist class – would likely be far more dangerous. And James implored us to hold in our minds the role shame and stigma play in minimising or dismissing

victims and survivors. I aimed to draw back the curtain on the
Trauma Industrial Complex, exposing its mechanics, urging
you to take a more critical view of the spectacle. Now, perhaps mischievously, I'm going to pull the rug out from under
you rather like trauma did me when I started this book. What
follows in Part 2 is an exhibition – a demonstration of how
overpowering a story can be even when you think you're wise
to it. In truth, this process is already underway and began at
the start of this chapter. You can feel it, surely. Some force
drawing you further into a narrative. So often, facilitators and
consumers of our stories come to them with a subtle sense
of superiority, as if we storytellers are in some way too vulnerable, require handling with kid gloves, or do not truly
understand the fullness of our stories' implications. This is due
to the power differential that exists between professionals and
people with direct experience – they call us 'experts', but not
all of them really believe that. That's why our narratives are
so often micro-managed within an inch of their lives and then
pumped out to the beat of that familiar three-act rhythm. How
about a narrative over which the storyteller has compete control; a story where the human being who lived it is the director
of the movie and not merely a facilitator's prop? If this idea
interests you, then you're in luck because that's what you're
all about to get: McGarvey has entered the chat.

In Part 2, I will audaciously deploy the very mechanics
we have only just deconstructed, slowly inducing you into a
state of narrative hypnosis, drawing you deeper into the story
I know best. My own. The story of my trauma, yes, but also
my experience of sharing it. The story of how I arrived at
the rather ironic and quite hypocritical position of urging you
to keep yours to yourself – at least until you gain a firmer

grasp of its meaning. While each person's story is unique, every tale shares a similar structure. Within this structure, the same developments drive our plots and characters forward. Every narrative contains similar themes. This is the entire basis of storytelling. The greatest stories endure because we all relate to them regardless of circumstances. If you have identified with any of this book so far, may it be in part due to the fact you see aspects of yourself in this Trauma Industrial Complex I describe? You sense vaguely that while the main thrust of your story checks out, certain details have been emphasised while others have been downplayed or edited out entirely. In your life, there are supporting characters who've served only to provide you with a foil to overcome or a foe to contrast yourself with – in your story their stories don't matter. There'll be others you've cast in roles that perhaps don't truly belong to them – the villains who let you down or stopped you from getting what you wanted. And there are surely stories you've told, as if they were true, the exact details of which you may never truly know. We all partake in these parlour tricks. As you proceed, I want you to try and hold in your mind everything we have so far explored – stories are often incomplete, may be told for self-serving reasons, and tend to portray the teller as the hero. As you descend my rabbit hole, notice how easy it is to forget all of that. Observe how effortlessly one may become engrossed in another's spooling narrative as I demonstrate the undeniable power and utility of highly personal stories while simultaneously deconstructing my own. In doing so, I hope to impress further upon you why stories matter, why their integrity must be protected amid their industrialisation, and why trauma's new salience presents not only risks but also unique opportunities. And for those of you who picked up this book in an airport or train station out of curiosity and not necessarily due to having skin in the trauma game, but who nonetheless sense your own narratives are incomplete, I offer you more than a grand tour of the Trauma Industrial Complex. I hope to model how you, too, might approach that long-dreaded rewrite you've been putting off for too long. Now, if you're all sitting comfortably, let the story commence.

PART 2

Processing

'The truth about our childhood is stored up in our body, and although we can repress it, we can never alter it. Our intellect can be deceived, our feelings manipulated, our perceptions confused, and our body tricked with medication. But someday, the body will present its bill.'

Alice Miller, *The Body Never Lies* (2004)

CONTENT ADVISORY

Part 2 contains vivid descriptions of traumatic events. Some readers may find these passages distressing. It also contains questions, observations and lines of enquiry about survivors' interpretations of their experiences which some may find inappropriate or unsettling. Certain trauma-related topics are explored in a manner some might feel unusual, uncomfortable or even offensive – proceed with caution. While every effort has been made to approach these topics with care and sensitivity, it is important to acknowledge that this material may evoke strong emotional responses. If you feel distressed at any point, please take a moment to step away, take a breath and remember you are only reading a book – some of these things actually happened to me. Engaging with trauma narratives can be difficult, and your safety and comfort matter to me – as does the truth.

Getting Your Story Straight

What stories do we tell ourselves about our lives, and are they all true?

The bookshop café is mercifully quiet on the last Friday before Christmas. I head instinctively to the back of the room, obeying an urge to remain distant. Aside from the anticipation of meeting someone I haven't seen in years, there's a calmness about me quite unusual for this time of year. I sip my second coffee of the morning, lost in my phone for a moment, and glance up to see a woman approaching my table – the one I've been expecting. Her name is Marlyn O'Connor, and I need her help, though not for the first time. Now retired, she spent her career as a child psychologist in Glasgow Psychological Service, supporting some of the most vulnerable children in different areas of Glasgow and at the Notre Dame Centre in the city's plush West End. In 2002, though, she had the dubious honour of having 17-year-old me referred to her for counselling. It's been a long time, but Marlyn looks exactly as I remember, and the years have been as kind to her as she was to me during our sessions all those years ago. I'm here to verify that trauma is indeed an important part of my story and not just an idea I've acquired from pop-psychology. If anyone can confirm my assumption, it's her. A month ago, she enthusiastically accepted my invitation to meet, and ever since I've been preparing questions – questions about trauma. The small talk flows easier than I'd expected, though I'm not sure why I ever doubted it would. After all, I've known Marlyn for more than half my life, having first met her the year after my mother succumbed to alcoholism. Back then, there were no smartphones to google your symptoms or speculate about your mental health. No online quizzes or clickbait screening tools. If you had a problem, you went to the doctor, and if you needed help, you got it in good time. And boy did I need help. With the pleasantries exchanged and two steaming mugs on the table, I begin by asking Marlyn what her first impressions of me were – if she can even recall. It's been over 20 years since we first met. While she still resembles the woman I remember, I've changed a lot. 'You were very articulate,' Marlyn recalls with a familiar grin, 'but you had no self-esteem.' I laugh – it sounds about right, though I haven't the heart to admit not much has changed in that department. 'I was struck by how willing you were to answer my questions,' she continues. By that time, my 'story' had already begun to take shape, both internally and through creative outlets like songwriting and a play I was working on. I'd aspired to be an actor back then, and with everything that had happened the previous year with my mum passing, accessing intense emotions came easily. In fact, part of my struggle was that my emotions often got the better of me – what the trauma field refers to as 'dysregula-tion'. Anxiety and anger caused me considerable distress, a fact Marlyn confirms as she recounts my GP's referral. 'You were struggling with anxiety and panic,' she says. 'You weren't sleeping well, felt you weren't coping with life, and mentioned you were using alcohol to help.' Her memory stuns me, inadvertently jogging my own. It was the year after my mother's death when I first encountered what would become lifelong nemeses: alcohol and drugs. By the time I met Marlyn, I'd already experienced – and begun craving – their pleasant effects, though I wouldn't cross into dependency for another few years. Living with my grandparents at the time, I was never in short supply of means to self-medicate. My grandfather was a drinker, and my grandmother due to ill health was reliant on prescription medication – both I believe carried traumatic wounds that they took to their graves. In our community, substances like alcohol and pills were as common as teabags and toilet roll. I was trying to finish my final year at school having begged to be allowed back in after leaving abruptly, feeling pressure to start making my own money. To say I enjoyed school would be a stretch, but I never detested it like so many of my peers. My struggle in school had less to do with the work itself and everything to do with the social dimension. At school, I stood out like a sore thumb. One report card noted my 'offbeat sense of humour' – a teacher's clumsy way of saying I didn't fit in. Everyone is, in some way, unique, but I was a little too different. At least, that's how I felt. The way I spoke, dressed, my interests and aspirations – they invited derision and conflict in various forms. This intensified when everyone in my year hit puberty and crippling teenage insecurities are masked by a practised insensitivity. All the boys seemed interested in sex, but all I wanted was to fall in love. I wanted to feel loved by someone. I acquired the idea that someone else's acceptance, affection and loyalty would relieve me of my torment.

95

Unfortunately, puberty for me brought with it not just a bad attitude but a severe case of chronic acne which plagued me for around two years; dozens of hideous bright red spots took up a residency on my face and neck, deforming my appearance with unsightly blotches, agonising boils and scarring. Whatever remained of my self-esteem by that point was completely shattered. There was nowhere to hide. I went from feeling like something of an oddball to certain I was a freak. Girls I fancied and had a strong sense fancied me, too, were embarrassed to go out with me, or even be seen with me. One even told me in the dinner hall queue that her friends felt she was too good-looking for me. Despite the wide range of emotions at play in this torturous period, my overriding day-to-day feeling was one of heartache. I was ugly and weird, consumed by thoughts of altering my appearance, my personality and even my beliefs, to render myself more acceptable to others. I had not yet gained the insight to understand the fundamental problem: I was operating from a deep pain inflicted in my early years – what many in the trauma field call 'core wounds'. If untreated, these wounds may lead to the formation of beliefs which over time become lenses through which you see yourself and the world. My belief was that I wasn't good enough. While certainly not as outwardly dysfunctional as many of the other kids I encountered occasionally at the Notre Dame Centre, Marlyn had her work cut out the day I rolled up. A willingness to talk openly and some semblance of self-awareness may have impressed her in the beginning, but my articulacy and seeming openness were, themselves, adaptations undergone over time, which helped me to avoid truly experiencing my feelings. Holding the world in my analytical mind kept it away from my heart. I grew adept at bamboozling people with words and sentiment, possibly as a means of keeping them at a safe distance, but at the same time I craved closeness and intimacy. In essence, characteristics many around me to this day regard as enviable qualities – intellect, conviction, humour, emotional intelligence – became decoys, creating the impression of a person who knew himself, and something of life, when in truth, I'd do anything but sit in the pain of the truth – even if the truth was as banal as being a little bored.

Drawing from her recollection of our early sessions, Marlyn tells me that I had experienced – or witnessed – several events that could be described as 'potentially traumatic'. It says a lot about the caution exercised by trained therapists that my experiences could be described as such – some were extreme to say the least – and even more about the ease with which the word 'trauma' flies of the tongues of online content creators in an age of self-diagnosis. My traumatic experiences likely explained my anxiety and anger, as well as my anxious attachment style, where close relationships are managed on the assumption that rejection and abandonment are imminent. School was a daily gauntlet, but Thursday afternoons with Marlyn were a reprieve. Despite my vivid recollection of my circumstances at the time, and my sense of connection to that narrative, I put to Marlyn that trauma is not the story of what happened, but rather it is the wound felt by the nervous system. A wound we often attempt to make sense of by creating a narrative around it. 'That's why the trauma therapy I used with you was different from other approaches,' she tells me, 'because you don't talk to trauma, you experience it.' Marlyn's passion for the subject is evident as she lights up at the opportunity to elaborate. 'It's a multisensory experience, involving sights, sounds and smells, all locked in when you've been traumatised. Unprocessed trauma can be triggered at any time by a sound, a smell or even someone saying the wrong word. It stays in your working memory, your right brain, making the past feel like the present. It lurks there, ready to surface, no matter how much you try to push it down.'

After our initial session in 2002, Marlyn took a brief history of my upbringing and experiences and, with my agreement, we began EMDR therapy – a treatment for trauma with a fascinating story of its own. EMDR, or eye movement desensitisation and reprocessing, came into being quite by accident in 1987. Its origins can be traced back to a chance observation by Francine Shapiro, an American psychologist. While walking in a park, Shapiro noticed that certain distressing thoughts that had been troubling her seemed to lose their emotional intensity when her eyes moved back and forth in a particular way. Intrigued, she began to experiment, initially on herself, to understand whether this effect was simply a coincidence or something more significant. Shapiro's curiosity led her to try the technique with others, focusing on individuals who had experienced trauma. She discovered that by guiding a person to move their eyes in a rhythmic side-to-side motion while recalling a painful memory, the distress associated with the memory appeared to diminish. Her early experiments were informal but promising, and it wasn't long before she began studying this phenomenon in a structured way. The results of her first formal trials with trauma survivors, many of whom struggled with debilitating symptoms of post-traumatic stress, were striking. Participants reported a marked reduction in their emotional reactions to traumatic memories after only a few sessions. This method, initially called eye movement desensitisation (EMD), seemed to hold transformative potential for people trapped in cycles of pain and anxiety. As Shapiro refined her technique, she expanded its scope to incorporate additional elements that would deepen its effectiveness. These included bilateral stimulation beyond just eye movements, such as tapping or auditory tones, and a structured protocol to guide the therapeutic process. The approach evolved into what is now known as EMDR, adding 'reprocessing' to reflect the method's aim not just to reduce distress but to help individuals integrate traumatic memories more adaptively into their broader life experiences. EMDR gained traction quickly, with trauma therapists drawn to its ability to help clients access and process deeply buried emotional wounds in a safe and controlled environment. The 1990s were a turning point for EMDR. Shapiro's work began to attract significant scientific attention, and researchers subjected the therapy to rigorous clinical trials. What emerged was compelling evidence of EMDR's efficacy, particularly in treating PTSD. This led to its endorsement by major organisations, including the American Psychological Association and the World Health Organization, as an evidence-based treatment for trauma. In the years since, EMDR has become one of the most studied and widely used therapies for trauma and related conditions, offering hope to millions around the world. Although its precise mechanisms are not yet fully understood, there are compelling theories about why EMDR works. One explanation is that the bilateral stimulation employed in EMDR – whether through eye movements, tapping or auditory cues – helps to 'unlock' traumatic memories stored in the brain and facilitates their reprocessing in a way that reduces their emotional intensity. Another theory suggests that EMDR promotes communication between the emotional and rational parts of the brain, fostering a sense of resolution and calm.

When Marlyn reinforced the idea that my memories didn't have to be so painful and that I could, if I did the work, begin to see them in a new and less troubling light, I simply believed her. For me, EMDR's success lay not simply in the treatment itself but who delivered it and the quality of the therapeutic relationship. While Marlyn observed at all times the necessary professional boundaries required, it was also palpable to me how much she cared about helping young people and how invested in my healing journey she really was. These are details that a wounded nervous system eventually detects and learns to trust – the essence of healing. While I am not qualified in the least to assert anything but my own opinion, I suspect EMDR succeeds when there exists some rapport or mutual understanding between the therapist and the client. Whatever the exact explanation, the results are clear: countless individuals report profound relief from the grip of trauma after undergoing EMDR therapy. What makes EMDR especially unique is its focus on confronting, rather than avoiding, painful memories. This goes against the grain of much of the contemporary discussion around trauma, which often emphasises the need to avoid triggers and protect oneself from distress. In EMDR, the process of healing involves deliberately bringing traumatic experiences into conscious awareness under the guidance of a skilled therapist, in a way that feels safe and manageable. The therapy's structured approach ensures that individuals can face even their most harrowing memories without becoming overwhelmed. Today, EMDR is a cornerstone treatment for PTSD and is also used for a variety of other mental health conditions, including anxiety, depression and phobias. From its serendipitous beginnings in a park to its current status as a globally recognised therapy, Francine Shapiro's discovery has reshaped how we understand and treat the lingering effects of trauma – and given new muscularity to the cliché that going for a walk when you feel down is not a bad idea.

The café grows busier as we talk. Our conversation feels less like an interview and more like an overdue reunion. I've always hoped I'd run into Marlyn again. For many years, I resisted trying to find her out of fear I'd learn she was dead. Now, as we chat with the familiarity of old friends, I feel a swell of affection and a childish part of me, if I'm honest, wishes I could just keep her. Marlyn was such a lifeline to me during a painful, confusing time in my life. Despite the weighty nature of our sessions, I always looked forward to them. She gave me permission to be myself – a rare gift for a self-conscious kid like me. Our sessions often began with me describing how I was feeling or what was happening in my life. I was in my first serious relationship, though it was already showing signs of trouble. We both carried wounds, expressed in ways that triggered one another – the classic anxious/avoidant dynamic described in attachment theory which has become popularised in recent years. A theory which I have found immensely useful later in life. Falling for anyone who gave me attention was easy; seeing when they weren't right for me was harder. The thought of breaking up with someone, of going on without them, and they without me on to meet someone new, led me to stay in relationships which wreaked havoc on my nervous system. Marlyn assured me that one day, with healing, I might find a suitable partner. We'd discuss briefly my current trials and tribulations, and how those difficulties stemmed from deeper wounds or unmet needs. Marlyn often reframed events and relationships in ways I found helpful, and it was during my time with her I began developing an awareness that I had trouble perceiving things as they were, due to my preoccupation with stories. Before each session ended, we turned to EMDR.

In preparation, Marlyn took a detailed developmental history. Normally, much of this information would come from a parent or carer, but in my case – my mother deceased and my father estranged – it was left to me to recount my early life. That in itself can create real difficulty for the professional child psychologist. The most severe memory we processed involved an incident when I was five. My mother, drunk and enraged, pursued me with a knife after I refused to go to bed. I ran to my room, and as she charged up the stairs, my father intervened, wrestling the blade from her. I was whisked away by someone – possibly one of her friends – in what felt like an emergency evacuation. Marlyn chose to process this memory last, instead beginning with less distressing episodes to build my resilience. While EMDR worked a treat for me, like most trauma-related topics, it draws its share of criticism. Coming under scrutiny in recent years, with critics questioning both its scientific foundations and the qualifications of some practitioners offering the service, advocates of the treatment have scrambled to ward off the naysayers. Sceptics argue that EMDR may lack a strong scientific basis, suggesting its core theories cannot be reliably tested and that its benefits might come from the general therapeutic process rather than the specific technique of eye movements. Concerns also arise from the growing number of individuals providing EMDR without full certification – a direct consequence of the rising demand for trauma therapy driven by widespread socioeconomic dysfunction and a booming but poorly regulated online marketplace. These issues emphasise the importance of ensuring that trauma therapies are provided by properly trained and certified practitioners to maintain their safety and effectiveness.

In my sessions with Marlyn I'd sit upright, palms up on my knees, while she tapped two fingers on the back of each of my hands. This is known as bilateral stimulation. 'As more events were processed, you reported feeling less anxious and angry,' Marlyn recalls. Each week, she asked me to rate my distress on a scale from zero to ten. By the end of therapy, I often reported zero, meaning the memory no longer caused me significant distress. For me, still largely oblivious to the mechanics of EMDR, my overriding sense is that the healing lay partly in feeling seen and heard at a time when I felt I had been discarded by family and left to process grief and abandonment on my own. While I was never short of a shoulder to cry on or a cosigner to vent my spleen to, few people around me possessed the psychological or emotional tools to offer precise guidance as to how a life should be lived. Marlyn was the first person to truly introduce me to the concept of recovery; the seemingly simple notion that my life was potentially more than whatever story I'd cobbled together was revelatory. She was the first person to truly witness the extent of my suffering and that alone had a soothing effect. She helped me draw a mental boundary between my thoughts and my feelings – until then I never understood the distinction. Despite all my years of carrying trauma and, indeed, mistaking its effects for innate personality traits, Marlyn may have been the first person in my life who possessed the necessary knowledge to truly help me begin tending to my wounds.

Part of the EMDR process involves implanting a safe space – a mental visualisation of a place you associate with safety – and a future scenario – a new story. 'We closed down the sessions, and as we always do with EMDR, we finished with the positive future scenario,' Marlyn reminds me. 'I said to you, "Right, you'll make up a little narrative about what you'd like to do in the future – in five or ten years." Then I talked about it, repeating the narrative, and did the tapping, saying, "Now that's it – it's installed. You don't even have to think about it." All the things you said you wanted – to be creative, to be in media, to have a family, to have children – those were embedded. When you asked me about it later, I thought, *You're doing it*. To me, that's great; it shows it worked. It's an unconscious thing, embedded in someone's mind, and they naturally move towards it. You're a perfect example of how that piece of work – the positive future scenario – can be effective.' Upon being reminded of this, I experience a gratitude chased by a mild discomfort. In hindsight, Marlyn's recollection is correct but my version of events differed slightly. My memory was that my work with Marlyn was not completed, for reasons I could not recall. This may seem a minor discrepancy, but the average self-authored life story is littered with these little deviations from the facts. One deviation may lead to another, and then another, and before you know it, you've effectively created a branch-timeline – an alternate reality which bears little resemblance to what really happened. We do this all the time when we tell the stories of our relationship breakdowns, tales of unrequited love or the glittering, high-flying careers we deserved which failed to materialise. As stated previously, most people get away with these little white lies, but what of those who facilitate, consume or tell stories? Is it not safe to assume that without proper insight into the facts, many of us fill in the blanks ourselves? If I recall correctly, the story I ran about my work with Marlyn was that due to our sessions ending abruptly (I aged out of eligibility) my treatment was somehow botched and as a result, I was in some way mentally defective, like a poorly wired appliance. Many years after my last meeting with Marlyn, having crossed into full-blown alcoholism, this plot deviation was developed further when it was revealed to me in the disclosure of a loved one that a once dear friend (whose home I'd built my mental safe space around in treatment) wasn't quite as safe as I believed. This revelation was itself traumatic. The idea that my wellbeing had been fundamentally corrupted gained new salience, offering what I felt at the time was a plausible explanation for my increasingly destructive behaviour. That was the last Christmas before I finally gave sobriety a serious attempt – trauma and alcohol are terrible bedfellows – as I went on to lose my job, my partner and eventually my grip on reality itself. When I stopped drinking, I did a great deal of work on straightening the story out and embarking on the repair work required to make amends for the mess I'd created, but at some point, as the novelty of recovery wore off, I grew tired of introspection and the blinding light of the truth, settling for surface-level self-awareness that passes for insight rather than deeper inner work – a deal most people cut in the end.

As we say our goodbyes and promise to meet again, part of me wants to cry out to Marlyn that I am not doing as well as she might think or hope I am. My profound sense that some central part of me remains broken almost finds expression in words, for who better to share this burden with than her? But I can't do it. She seems so pleased to see me looking and sounding so happy, and together. At once I'm struck by how much time has passed, how the world around us is unrecognisable to the one we met in all those years ago – and how much has changed yet remained the same.

*

I come away from the reunion feeling uplifted and also fearful. It was pleasant to spend time with someone who truly knows me, but I leave with a strong sense that the book I thought I was working on may not be the one that needs to be written. I went in with a game plan to establish the truth about my own level of trauma – this was thankfully a success – but I was also keen to loop Marlyn into the narrative I was creating around the harms of trauma discourse. At various points in our conversation, I found myself speculating as to what Marlyn might think of the current trauma landscape: the misapplied terminology, the cowboy online counselling and the rise of public disclosure. I thought to broach these topics with the hope of eliciting some validatory responses which reinforced my own sceptical intuitions, but any attempt to draw her was gracefully parried. While I'm sure Marlyn is well aware of the eruption of interest and debate around trauma and has her opinions on it, her lack of interest in taking an antagonistic position was echoed by Suzanne and James a few months later. Sure, she acknowledged the dangers of unregulated therapies and the risk to public health when perverse financial incentives run amok, but like Suzanne and James, she didn't appear in any way perturbed by the notion that more people than ever before are using the language of trauma in their everyday lives. Her concern, like Suzanne and James, is that more people need help who aren't getting it. Are they all missing a trick? Or am I trying to shoehorn them into my own little subplot. While I believe a tough conversation about trauma is necessary, my time with Marlyn as well as with Suzanne and James leaves me with one overriding thought: could I be approaching this topic the wrong way? Unlike me, they each seem resistant to be drawn into contentious discussions which pit various forms of trauma or treatment against others. They have their beliefs and seem in their own ways resolved that making those values visible is enough – criticising or attempting to discredit others isn't part of their repertoire. Is this part of a strategy or is this a reflection of who they are or, rather, of who they've worked hard to become. Setting my pride aside, I must confess this causes me to pause for thought. The Trauma Industrial Complex indeed poses risks – to those sharing their stories and to the integrity of this distinct form of storytelling itself – of this I am certain. Yet, my irritation – bordering on resentment – at what I perceive as the contagion of trauma as a concept may in fact be symptomatic of the very problem I believe I am trying to confront. I want this story to be true, but is it? What bothers me about the recent salience of trauma? Is it really concern for the safety of others? Am I harbouring a strain of exceptionalism? Despite my working-class image, is there an element of snobbery at play here? Do I believe that only stories like mine, told my way, truly matter? Perhaps some inverse class-prejudice arising from the assumption that the discussion is dominated by affluent professionals or middle-class millennials desperate for a hard-luck story? Could I be attempting to gatekeep this culture much like music purists guard jealously their obscure genres and pristine vinyl collections, insisting that experiences only have true value when leveraged for social change or to defend the vulnerable? If my ultimate goal is to foster a culture where stories are told safely and for reasons beyond mere catharsis, does harsh, sneering criticism serve that purpose or undermine it? At this pivotal moment – when trauma is finally part of a paradigm-shifting cultural conversation empowering countless previously silenced voices – do I really want to sow seeds of doubt? Part of me does, to be sure. The same part of me that comes alive in the heat of conflict. That part of me that would rather self-exclude in advance of being rejected – the essence of my lone-wolf campaigning style. Admitting this is difficult, but in the spirit of this book – where I'm asking readers to think harder and be uncomfortably honest – it feels appropriate. Suzanne, James and Marlyn model what strikes me as a healthier approach. By being kind and patient, even amid disagreement, they gently push me to reflect. Their softer tones and flexible thinking beg me to reassess not only the issues but also my underlying motives. They compel me to reconsider the story I tell about my own trauma and the broader narrative I construct around the culture of storytelling itself. Sitting with them, aware of their integrity and expertise as they gently challenged some of my analysis, I began to feel I may have missed something during my solitary descent down the trauma rabbit hole. With all my talk about getting the particulars of your story in order, and understanding the intent behind telling it, it seems I'm still as capable and willing as anyone I may criticise to turn a blind eye to some of the finer plot points. I can bamboozle packed theatres and conference halls with my stories, tie experts in knots with my verbiage. But what is all this awareness, when I'm made so uncomfortable simply being in the presence of those who to my eye appear healed? What am I afraid they might find out about me? Why do I feel like a fraud who does not deserve to be here. Do you really think I'm the only prominent survivor of childhood trauma who talks a good game in the public eye while privately suffering behind the curtain? How many of the Trauma Industrial Complex's leading lights are actually working the principles they preach? I wish I could say my mouth has been matching my feet of late, but as James used to tell me whenever I'd hit rough patches in recovery, I'm often 'trying to ride two horses with one arse'. Sure, it's been years since my last bender, but beyond doing the bare minimum to stay sober, I haven't exactly flourished. Certain coping strategies remain integral to my survival and even my livelihood, or at least that's how it feels. My life remains littered with stubborn attachments I know I'd be better letting go.

This public-facing persona is only part of the story. In this world of high-minded debate, tribal rancour and righteous fury, most of us lead double lives of one sort or another. And none more than the very people so desperate to be seen as 'authentic'. We'll touch on the notion of authenticity later in our journey. For now, I'll just say that as someone who is frequently referred to as 'authentic', leading my own double life has become physically and mentally exhausting. This introspection feels like the tedious self-examination many of us on healing journeys must undertake – reflecting on our attitudes, thoughts and behaviours periodically to gain deeper self-awareness. It's not for everyone, and for those unburdened by trauma or addiction, it remains entirely optional. But for many of us, this inner work, tiring as it can be, is essential. It's inner work I've been slacking on lately, if I'm honest. My experience tells me that an unexamined life risks relapse, mental health decline, and relational turmoil. This work, though difficult, has been for me nothing short of lifesaving. The problem is it never ends – that's why so many of us don't do it. Much of the discussion around trauma irritates me because beneath its excesses lie home truths I have been unable to face. I am not in a process of healing; my recovery has drawn to a halt. Ironically, this has happened partly because I have become a media personality who has grown successful enough that I can distract myself from the reality of my own health by hopping onto my soapbox and analysing everyone else's. Then there's the labyrinthine Trauma Industrial Complex, doing its best to keep me picking from that all you can eat buffet of neurodivergences, attachment theories, personality disorders and trauma responses. Poring over it all endlessly has become a poor substitute for consistent action. Action that my experience tells me works. I could talk for days about trauma and even give the impression I know a thing or two, but this circular preoccupation has, I fear, become the latest distraction from pulling myself together again and getting on with the graft. And I'm not the only one.

'People who have survived atrocities
often tell their stories in a highly emotional,
contradictory and fragmented manner.'

Judith Lewis Herman,
Trauma and Recovery (1992)

CHAPTER 7

Main Character Energy

*What is your origin story,
and does it help or hinder?*

It was the day a little white envelope landed on the living room floor, tossed there by my six-year-old daughter hiding out of view at the bottom of the stairs, that I finally accepted I was broken. 'Dady,' it read, in her best attempt yet at handwriting, 'I sora yu are havin a hard time. Dady, I luv you.' Had I felt less ashamed of the fact that my own child was attempting to emotionally support me, I might have wept uncontrollably. But the tears – they just didn't come. It's not that I don't feel emotions – it's that I can't seem to express them when I really need to. Instead, I rely on words to describe feelings, which – even for me – have firm limitations. Since I put down the bottle over a decade ago, I've found it hard to cry, even in times of real sadness. Friends' funerals, family feuds and even genuine heartache have failed to produce as much as a drop. Where does my pain go? The hardest I've wept, ironically, was during short relapses over the last ten or so years of recovery from alcoholism. The day I collapsed in Central Station, Glasgow, after a heavy night of drink and drugs, I asked the police officer to cover my face so I wouldn't be recognised while we waited for the paramedics. In that moment, I thought of my son and how embarrassed he might be if word got out that I'd got myself into such a state. As the officer kindly pulled my hoodie over my head, complaining to his colleague about the outrageously long wait for an ambulance, I remember breaking down so hard emotionally that I almost couldn't breathe. For all the gifts I've been given in this second act of my life – a sober, married father with a successful career and friends and family who are there when I need them – something is still very wrong with the picture. I've become inured by ruminating and speculating about what's wrong with me, my wild imagination stoked by the omnipresent Trauma Industrial Complex. This apathy is never wise, but it's risky business when you have children. All the labels, frameworks and treatments I could recite offhand, and yet the feeling that something terrible is about to happen rarely leaves me for long. Whatever public face I portray, privately I am always preparing for the final calamity. I feel, in daily life, for long stretches of most days, the way I imagine fearful flyers must 30,000 feet above the earth. And to put my spin on a well-worn turn of phrase, for me, turbulence is other people. I can't be around anyone for too long without feeling the strong urge to leave, even just momentarily, to catch my proverbial breath. I don't go anywhere unless I have an exit-plan, and I become restless and agitated the longer I linger. My social instincts seem calibrated for a particular purpose. They seem set to detect, track, and anticipate, in every interaction I have, the intention, temperament and mood of the other person. What I need to know, before anything else can happen, is that I am safe and secure. I've been like this for as long as I can remember and it's hard not to wonder sometimes, how this preoccupation may be affecting my own kids. When holding them in your arms for the first time, the mere thought of them coming to harm is painful to consider. The instinct to protect them is so over-whelming. Regrettably, like so many content creators I have come to loathe, I too often struggle to see my children through any lens other than trauma. But unlike those commentators, as well as a protective impulse, my preoccupation also manifests as fear that I am going to hurt them. Sometimes, as they talk to me about their day, or share their thoughts and feelings on wildlife, the moon, their friends, I suffer intrusive visions of another possible future scenario. One implanted long before I ever met Marlyn. A scenario in which my children are sud-denly stricken with terror. I picture them running away from me as I angrily pursue them. Sometimes I even see them being physically hurt by me. In my mind's eye, they plead, desperate for comfort and consolation – from the person who abused them – and I refuse it. It is then that I am filled with a bizarre blend of relief and rage. Relief because I know in my heart I could never in my right mind harm them, and anger when I wonder how anyone could have done those things to me, when I was a child. Into the bargain, every now and then, for no reason at all, scenes from what feel like my past project themselves sharply onto the busy wall of my waking mind. These scenes do not follow any specific pattern. They are not triggered by any specific stimulus, and I rarely recall them voluntarily. Yet, when they do arise, it can feel very much like I'm always thinking about them. I couldn't even tell you which order the events fall on my internal timeline. All I can say is that for most of my life, these little movies have been playing in the dark recesses of what I assumed is my memory. Despite my feeling that almost all of them occurred before I was ten years old, their detail remains vivid. I could not say what year each event took place, nor what happened the days before or after; these movies exist only as contextless fragments connected by one thread – they all, I believe, happened to me.

I push a cat off a window ledge. It claws the wall outside. I am crying at the front door of a nursery as my favourite teacher explains that she is leaving. I don't want her to go. My mother screams as she rushes out the door and down the stairs, partially clothed, to retrieve a cat. We are now on a rare family walk. My mother and father are both present. This is what normal feels like. My mother lets a dog off the leash. It's hit by a joyrider on a motorbike and thrown into the air. The vehicle scrapes to a halt in a trail of sparks. Why do things like this always happen to us? There is a thunder and lightning storm. I am alone in a dark room. The window is open. I am now suspended outside, hanging upside down in the rain. I am now inside a cupboard with another child. A girl. Sunlight is breaking through the locked doors. I hear people outside laughing. Older kids. They won't let us out until we kiss. I am watching a black cat walking on a kitchen counter by an open window. My mother is crying. She is punching a man in a black motorcycle helmet, as a dog howls by the roadside. Now I am lying in the bottom bunk late at night. Other children are nearby. One is a baby. It's daytime. A man kicks a baby across the floor. I look out of the window of a high-rise. Another child is tied to a chair. I don't know who the man is. I can't do anything to help anyone. There is a tussle between adults. I am picked up and carried out of a house by a woman with brown skin, into a car and driven to another place where a painting of a man in a kilt, holding bagpipes, is mounted on the wall. My mother is in the garden. She is trying to dig up the dog buried there the day before. She is crying. She is asking for my help. I am torn. I want to help her. I want to make her happy. The neighbours are all out. We are not normal. My aunty stands at the top of the backdoor stairs, pleading for me to come inside the house. I am back in the dark room with other children in the bunk beds. I'm the eldest and I'm trying to keep them quiet. We need the man at the door to think we are not here. I arrive home after school. Charred furniture is piled up in the front garden. I am embarrassed because I have a friend with me. A teacher bangs her fist on a desk and screams. She makes me admit to something I did not do, in front of the class. I am in a phone box, explaining that my mum is in the toilet, bleeding. My granny puts the phone down and tells me there has been a fire at my house. My granny pulls me away from a woman with a melted face. My granny will protect me. My granny makes me feel safe. A terrapin swims in an old chip pan, in a dark, smoke-filled room. My mum is showing me how to throw a punch. She tells me to put my thumb inside my fist. A glass bottle shatters on the wall behind me as I duck behind a couch. An old woman with a melted face is laughing. A budgie is celebrated for being able to say bad words. My mother is annoyed because I won't go to bed. I am showing off in front of her friends. She walks into the kitchen and takes a knife from a drawer. My mum walks into my class-room to fight a teacher who frightened me. A fireball rises to the ceiling from a chip pan. We run out of a narrow kitchen and into a hall. This is not easy reading, is it? Some of you are starting to think I'm making heavier weather of this trauma malarkey than I need to. Some of you are beginning to feel very uncomfortable. This long passage recounting flashbacks of traumatic experiences is beginning to feel a little excessive, don't you think? Can't you feel it? So gratuitous in its detail, you now wonder whether you really are interested in trauma, or just those little stories those other people tell you about it.

Could

this

be

because

when traumatic experiences are disclosed publicly, they usually conform to that familiar

rhythm?

They follow structure which renders them palatable.

They conclude within a fashionable timeframe, filling your foolish hearts with hope that all

adversity can be overcome with just the right dose of RESILIENCE.

Well, I do apologise for any discomfort, but I will not wrap this up in a nice little TED Talky ribbon for you ... because the truth is my friends ...

that

is

not

HOW TRAUMA FUCKING WORKS.

I am walking across open ground between two high-rise flats in a storm to get cigarettes for my mum and a man and blown off my feet and across the grass while an angry teacher kicks open the door of the boys' dorm and threatens to kick the fuck out of all of us if we don't go to sleep. He is the same teacher I've seen attacking my friend and the same teacher the girls feel uncomfortable around. He is a teacher I am supposed to go to if I have problems at school and I do HAVE PROBLEMS.

All the other teachers are scared of him, too. Nobody stands up to him but one day I will. Adults are watching me from the window of a high-rise as I lie on the ground. I have so far to walk, and I am afraid to stand up in case I blow away. They are laughing at me. My mum is angry that I am frightened of a bully. She takes me to his house and makes me fight him in the street. This happens a few times or maybe just twice or was it just once? Other kids are watching us. A teacher asks me what happened to the jacket I was wearing when I blew away which got ripped that I am somehow wearing to school. An angry drunk man is screaming threats through a letter box in the middle of the night. He is kicking the door in. If he breaks through, I will not be able to protect the children. We are not safe here and nobody is coming to help us.

*

What you've just read is what many in the trauma field call 'oversharing' – an impulse to disclose personal information which may compromise the sharer in a manner they later regret. People who overshare, it is said, often lack healthy boundaries, because of traumatic experiences. I could be wrong, but I'm going to hazard a good old guess that I am one of those people. Having sat through my overshare, it may not surprise you that somewhere along the line, I acquired the idea my life was uniquely tragic. How or when this notion took shape, I cannot be sure. Was it the alcoholic mother whose unreliability plagued my early years? Or maybe it was the deprived community I was raised in, notable for the poverty, violence and addiction it so effortlessly produced. Perhaps leaving the family home abruptly at 16 before becoming an alcoholic in my early twenties was the genesis of this narrative. Or possibly the reactions of shock and sadness on other people's faces when I regaled them with detailed accounts of my difficulties did the job. Whichever way this idea took root, whether a sudden epiphany or a gradual view constructed over years, the belief that my upbringing was hard and had damaged me beyond repair, at some stage, became the controlling idea of my story. A story that was almost as hard to bear as the trauma itself. I've been telling some version of this story for over two decades and yet it still feels incomplete.

My mother's name was Sandra. She was a troubled woman who died in March 2001, succumbing to the alcoholism that had plagued most of her short life and the only reason so many people are aware she even existed is because I wrote about her in a book. Some even say I threw her under the bus. People who knew her well tell me she was caring and good humoured, though my memories often feel at odds with theirs. She too led a double life. By the time she left us – when I was around ten – I felt more relief than sadness at first. And when she died some years later, I didn't feel much at all until the hearse carrying her coffin pulled up behind me at her funeral. I'm not the first to lose a mother in the chaos of adolescence, but at the time, it felt utterly unique. If her death had been the only challenge I faced then, I might have weathered it and moved forward. But by 17, I was already on my own, having been asked to leave the family home before we learned she was sick. This was a painful time for my family. One that many of you will recognise, when unprocessed trauma triggered by stressful life events erupts through every member of a household, shutting everyone down emotionally at precisely the moment they should be drawing closer. Rather than pleading to be allowed to stay, I called my old man's bluff in the midst of a heated argument. Throwing some clothes and belongings hastily in a black bin bag, I walked out and never looked back. Struggling at school, I slept wherever I could: on friends' couches or with family members whose kindness only underscored my growing sense that I was becoming a burden. It was the kind of time when most people would turn to their mother – for comfort, advice or practical support. Instead, standing in the dusty loft of a friend's house where I slept, I was told she was in the hospital. Somehow, even that conversation became another argument. Before I even had the chance to visit her, she was dead. Hers wasn't a peaceful death either. You don't slip away gently when your liver fails. You turn yellow, skeletal, unable to eat or drink. Your legs and abdomen swell grotesquely. You writhe in agony, vomiting blood, delirious. If you're still aware, you beg for death. If you're lucky, drugs dull the pain enough that those around you might mistakenly believe you're at peace. She died as a direct result of a story she told herself for most of her adult life. A story about how she couldn't stop drinking. In truth, she probably could have. She probably wanted to. But that story, and its power over her, compelled her to continue consuming alcohol even after she was told by doctors it would kill her. Sometimes, I wonder if it was a blessing my siblings and I didn't see her in those final days. Yet I also wonder if a visit might have brought some closure – a satisfying end to that plot-thread of my story. Did she expect us to come? Did she wonder why we didn't? Even though she'd shown little consideration for others in her life, I sometimes think about what we might have said in those final moments. Could I have held her hand, told her I loved her, and meant it? Could I have forgiven her? Circumstances

robbed me of that chance, leaving so much unresolved – especially the questions I would have asked had she lived longer.
Why did you leave us? Why didn't you stay in touch? How
could you go about your business every day for years without
knowing where or how we were? Did you miss us? Did you
worry? I think I know the answers, but even now, at 40, their
open-ended nature still stings. As a parent myself, I look at
my children sleeping and can't fathom going more than a day
without contact. And yet, by 17, I was grieving and struggling
alone in a world I was quite unready for.

That's my origin story. The story I tell myself. Most who
live with trauma have one and I've been telling mine for 25
years. That story is probably the reason you're reading this
book. Despite my talents for writing, performing, for music
and public speaking, it's this story that catapulted me from relative cultural obscurity into the pages of the *New York Times*,
and onto the cover of the *Guardian*'s cultural supplement.
What you may find hard to believe, however, is that I had no
idea this one aspect of my story – my relationship with my
mother – would generate such widespread interest. The book
this story was told in was about so much more than that, but
readers and the media seemed most drawn to sections where
I detailed the traumatic events involving her. I retell it to you
now, partly to provide context for why I am writing this book,
and also partly because I feel that without it, people may lose
interest in anything else I have to say. For then, I might have
to sit with some feelings I would much rather avoid.

When I first shared the idea for this book with my dad, he
was intrigued by the concept of how we construct narratives
around our experiences. Like all fathers and sons, we've had
our ups and downs. Aware that there were aspects of my public

rise over the previous few years that he likely struggled with (concern for my health, fear of family business being in the public domain) I felt it important to make him aware that my next piece of work would be similarly personal. One minor consequence of the story of my mother becoming public in the way it did is that my dad's role in my life was inadvertently under-emphasised. Worse, some assumed he either wasn't there or that he played some active part in my adversity as a child – nothing could be further from the truth. That's just one of the many unforeseen consequences arising from my public storytelling. When I wrote my first book, it wasn't until the very end of the process that I began to reflect, perhaps for the first time, on what a beautiful and delicate man my father is. The story of his resilience as a single parent in the face of poverty, my mother's alcoholism and the shadow cast by his own difficult upbringing. Back then, it wasn't the done thing for a man to raise three kids on his own but that is exactly what he did. He didn't just raise us, he fought for us. He stood his ground in the face of the various other influences present in our lives that he feared may have led us astray. It was only when I matured in sobriety and became willing to reexamine the prevailing narrative that had defined my early adulthood, that I experienced compassion for his parental predicament: he was just a kid himself when I was born. He's been nothing but supportive of my career since it took off and so, naturally, I felt comfortable broaching the concept of this book with him, to hear his thoughts. I told him I planned to delve deeper into my memories of my mother. We talked about the night she came at me with the knife – a moment etched in my mind. But during this conversation, something unexpected happened: my dad's version of events differed from mine. In my memory, she barged into the room, pinned me against the wall and held the knife to my throat. I remember her eyes. I remember thinking I was going to die. I described my memory vividly in my first book. In that memory, my dad intervened, pulling her away, and someone else carried me to safety. But in my dad's recollection, she never got close to me. Everything else I remember was corroborated by my dad's recollection; I wouldn't go to bed, I was being cheeky, showing off, my mum went into the kitchen, opened the drawer, took out a knife and came after me. I ran into the hall and up the stairs into my bedroom – all of that we agree on. But in his story, she never reaches me, he got to me first. The knife was never within cutting distance, according to my old man. He even suggested that she'd intended the stunt as a poorly judged and drunken joke – a terrible one even by her standards. His account left me confused at first. If it had all been a joke, why did I run? Why did two adults rush to stop her? In truth, these questions came not from a desire to understand but to bolster my interpretation of the event. Despite my obvious reservations, it's not beyond the realm of possibility that my five-year-old self misunderstood something of that evening. The wound still occurred, of course, and even my dad's version of events, if true, is sufficient to create a lasting trauma; the breach of trust, the obliteration of safety, the terror that overwhelms your capacity to cope on your own. Even then, I became willing to accept the notion that some aspects of my origin story may have happened differently than I remember, and that the complete, objective truth may now be unrecoverable. It's a huge and crushing concession to make, especially in the public domain. By sharing this with you, please know I am not suggesting that every story of a traumatic incident told publicly may be a product of false memory. Believing victims is essential. In this particular case though, I was a young child, which means there is a chance that my recollection is imprecise. Then again, my old man's could be too. As difficult as it feels to accept that an important aspect of my public origin story may not be as true as I think, it's a possibility I am at least willing to accommodate. What matters here isn't the precise truth of the event, but the imprint that was left upon me and all of the conclusions I consciously and unconsciously internalised from that day onwards. In response to my dad's unwitting revelation, which I did not dispute nor seek further clarification, I came away from our trip down memory lane aware that two conflicting accounts of this pivotal plot point now existed. That may seem shocking, but this is not uncommon where stories of family dysfunction exist. Indeed, it's the rule rather than the exception where most stories about our lives are concerned. If my dad is right, then it reveals the true power of the stories we tell ourselves and how they may fundamentally alter the course of our lives. The difference between the usual stories and stories about trauma which receive a level of prominence should be obvious, however. What risks are posed by a story told that may not be wholly true? When your identity is tied to a story, questioning its veracity can feel like a betrayal. But is it? I'm not so sure. When I refer to the truth in this context, I'm not just talking about the specific details of what happened. Part of our story may be the intent we ascribe to others, or their level of culpability. I thought my mum wanted to kill me, and this belief became seared into my bones. It led me on a wild-goose chase in life, craving affection and acceptance, often from people too wounded themselves to provide it. While countless traumas arise from the calculated actions of malevolent perpetrators, just as many wounds occur as a consequence of caregivers not knowing any better at the time. This should never be given as an excuse for abuse or neglect, of course, but there may come a time many years after a wound was inflicted, when we view someone who hurt us in a gentler light. We may regret the story we ran with in our minds or feel a desire to redraft it to reflect new and important complexities. For some of us, however, who go public without careful consideration, the first hurdle we may face when attempting to rewrite our story is pride. In the rewrite, we may have to confront the fact that while our story still rings largely true, some finer points require adjustment. I fear many, operating without guidance, may come to cling to their stories due to the sense of safety they provide. In the long run, however, retaining their incomplete narratives will offer diminishing returns. Over the years, I've been asked for advice by hundreds of victims, survivors, aspiring speakers and writers. They want to know how to tell their story – how to inspire others, to transform pain into power. Many seek the life they think I have: a meaningful, sometimes lucrative career built, in part, on personal storytelling. But here's the hard truth I've stopped sharing with them because I suspect few really want to hear it, given how hard it goes against the grain of our affirm-only ethos: if you believe your story is important enough to tell the world, and that doing so will heal you or inspire change, the first thing you might want to do is make absolutely sure that story is true.

'To suffer without speaking is the mark
of a hero. But suffering that is spoken
can be turned into spectacle.'

Roland Barthes, *Camera Lucida* (1980)

CHAPTER 8

The Story We Sell

*What toll does public visibility
take on those who share
their adversity publicly?*

The Glasgow Film Theatre is packed – an unusual sight, and the first of several peculiarities I'll note on this particular evening. I'm here as a guest on a post-screening panel, bracing myself for potential questions from a host and audience I assume to be predominantly middle class. Mercifully, as the house lights dim and the feature begins, my nerves recede. The film opens with the pages of a notebook, revealing immaculately handwritten questions: Can you describe your parents? Their career choices? What are your earliest memories? The name Helen is scrawled at the top. As the pages turn, more questions emerge: What music do you like? Can you sing? Why have you kept your distance from me? The final query, though – Are you maternal? – lands like a punch to the gut. These are the questions of a child, nervous and yearning, directed at a mother who abandoned her long ago. That child's name is Lisa, her mother, Helen. *Blue Bag Life*, a poignant and introspective documentary by British artist Lisa Selby, delves into the lifelong impact of addiction, grief and generational trauma – realities often glossed over in media portrayals of family breakdown. Through an intimate tapestry of video recordings, photographs and personal artefacts, Selby explores her complex relationship with her troubled mother. Despite Helen's years of neglect and substance abuse, Selby paints her with surprising empathy, avoiding the clichés and stereotypes that often plague stories of hardship. Raised on a council estate by her dad, Lisa weaves together past and present while living in relative poverty, pursuing her dream of becoming a professional artist while acting as an unofficial and unpaid carer for her partner, who is himself navigating addiction. The film's title is drawn from the blue bags associated with drug use, a poignant symbol of the cycles of addiction and family breakdown that have shaped her life. *Blue Bag Life* stands out in an oversaturated genre. Eschewing sentimentality, it opts for raw candour and emotional complexity. The film adopts a collage-like format, mirroring the unpredictability of trauma and healing. This approach resonated strongly with audiences, earning it the Audience Award at the London Film Festival in 2022 – and, later, a BAFTA nomination for Outstanding Debut by a British Writer, Director or Producer. Yet, at this Glasgow premiere, well before the film's critical acclaim, Lisa and her collaborator, director Rebecca Lloyd-Evans, are understandably anxious. Their triumph, while assured to me, remains uncertain to them.

As someone with my own lived experience credentials – books, documentaries and live shows reflecting poverty and trauma – I feel a deep connection to their journey. I've stood where they stand: terrified that a long-held dream might soon come true, unsure if success or failure is the greater fear.

I understand the world they are stepping into – a world eager to commodify trauma as the latest accessory for third-sector organisations and social commentary. When the credits roll to heartfelt applause and the house lights rise, I scan the predictable sea of misty-eyed faces. Lisa, Rebecca and I are introduced and invited onstage. Before the discussion begins, the host issues a trigger warning – the second peculiarity of the evening. It may seem bizarre given I included one in this book, but trigger warnings bug me. An odd stance, perhaps, for someone who has sought trauma treatment multiple times but that's precisely why they irritate me so intensely. I may be wrong, but they rarely feel like they're being offered for the benefit of people with trauma. Instead, they appear to be for the benefit of the person or organisation issuing them. They have become, in too many cases, a means of telegraphing one's values or politics. I detect something rather performative and impractical in the way they are often dispensed, particularly in the arts. I'm not opposed to them – content advisories have existed in various forms for decades – but in many cases, they elicit either unnecessary vigilance in those who may otherwise remain calm in their absence or they display a weak grasp on the fundamentals of trauma itself. PTSD is not a mild discomfort or a passing distress. It's debilitating: the floods of emotion, violent shaking, sleepless nights and relentless internal threat-modelling that leaves you hollow. For all the good intentions behind these warnings, they often misunderstand the nature of triggers. Triggers are random, unpredictable. They're impossible to pre-empt entirely, which is part of their terror. Proper trauma treatment emphasises confronting and gradually learning to trust your ability to cope with discomfort, rather than avoiding it. Take the blue plastic bag, for instance – the film's quiet, recurring symbol. For Lisa, it was never needles that triggered her. It was those bags: everywhere and unavoidable. Caught in trees, trampled into gutters. But instead of shrinking from them, she found a way to confront and reimagine them – through her Instagram account, through art, through the film itself. The bag, once a source of pain, is now embedded in something she's shaped and made meaningful. The words trigger warning don't reassure me – they agitate. They prime me to scan for problems or threats, activating my negativity bias. So, on this evening, it may be worth asking: who is this warning really for? Certainly not the deeply vulnerable. Anyone with severe trauma is unlikely to put themselves in a situation they're unready to handle – their body wouldn't let them. Moreover, anyone sitting through *Blue Bag Life* has already endured its emotional weight; the warning comes far too late if safety is the concern. Given Lisa's own experience of trauma of various kinds, surely, she'd have thought to preface her own film with such a warning if she felt it may cause viewers distress – she didn't. That said, as I bristle with agitation, I must remember that my experience is mine alone. I remind myself – again, as much as it pains me – that not everyone sees this world the way I do. Not everyone *feels* it the same way I do. That's the danger in conflating shared language with shared experience. In truth, what annoys me is not the notion of people being overly delicate, but that these warnings often serve organisations eager to flaunt their 'trauma-informed' credentials for cultural clout. It's not that I object to care – I object to performance masquerading as care. Trigger warnings in the arts, in many cases, cater less to those living with trauma and more to those who have adopted a particular narrative of their own fragility based on ideas they've acquired online. Ironically, this performative sensitivity often betrays a lack of genuine insight into trauma and contradicts how often vulnerable people are treated within these industries when no one is looking. Behind the polished façade of this premiere likely lies an unglamorous reality. Here at the GFT, for example, staff – ushers, technicians, cleaners – are currently trying to unionise for better pay and conditions, I imagine they will meet at least some resistance, and in my experience that resistance sometimes poses as a paragon of progressiveness. It speaks to the window-dressing trauma-informed jargon often provides, like the kind seen in so many cultural venues, draping themselves in the language of inclusion while pricing out the working class or excluding disabled audiences through inaccessible spaces. It's hard not to notice the disconnect between the rhetoric of inclusion and the reality of access, between the performance of care and its actual practice. Tonight, Lisa and Rebecca, I suspect, are exhausted. They're managing the usual chaos that comes with being on the road: late trains, missing taxis, suitcases dragged through unfamiliar streets. I'd been able to cover their accommodation myself after learning that a hotel hadn't yet been arranged when I invited them to join me on my podcast the morning after the premiere. At the time, I saw this as another example of how lived experience workers – artists included – are often left to patch the holes in under-resourced projects. While my instincts are rarely wrong in this regard, Lisa assured me I was wide of the mark on that occasion and I was more than happy to stand corrected. In truth, she and her collaborators had been carrying the costs of the tour themselves, directing every bit of funding into travel, accommodation and simply getting the film out into the world. No one was making a profit. It wasn't about personal gain. It was about getting the story seen, heard, felt. And that hustle – that urgency – is familiar to anyone who's ever tried to raise awareness when the budget runs low, yet the mission remains. That enthusiasm is what powers the lived experience juggernaut. We keep going. We stretch. We make painful compromises. You're 'on one', as Lisa put it. But even with Lisa's less cynical take on some of these burdens, I suspect part of her pragmatism arises from necessity. Like so many of us with first-hand experience of adversity, who choose to tell our story, she's likely juggling the invisible crises unfolding back home without showing it. Because we learn, quickly, to play nice. Not to be too difficult. Not to seem ungrateful. We learn to make ourselves smaller. Because being seen as needy or disagreeable can mean opportunities being withdrawn. None of this, of course, is visible to the audience. And that's by design. What they want is a tidy arc: how bad it got, what we learned, where we are now. A neat beginning, middle and end that ruffles few feathers. Nothing too messy. Nothing too unresolved. Half the time, trauma survivors possess little frame of reference for what being treated well or fairly even looks or feels like. As a result, on the rare occasions our basic needs are met, we may feel extremely endeared and become fiercely loyal – only to realise later we'd been love-bombed.

While the front-facing spectacle emphasises inclusion, diversity and care for the vulnerable, the people with the so-called lived experience – Lisa and me – are the true labourers of the night. And yes, the room is very white and very middle class, but I've come to appreciate that's just another aspect of the labour. Lisa and I have spoken before about the importance of cross-class collaboration – how mutual respect and support across difference might be the only way we begin to untangle some of this. But such collaboration hinges on those of us with lived experience intuitively navigating the minefield of middle-class assumption, entitlement and passive aggression. Not all people with first-hand experience will enjoy the same treatment or respect as Lisa and me. In truth, we represent a slender minority. The story I hear time and time again, behind the scenes, from lived experience workers and storytellers when they feel safe to speak freely, is that they often shoulder an unseen emotional burden. They face invasive questions and are often paid poorly – if at all. They feel too often like props being wheeled in and out of view, at the behest of facilitators. The audience walks away feeling enlightened and inspired, but for those doing the telling, the cost can be immense. We're expected to embody resilience while performing emotional labour for others. To assert our own needs risks rejection – another kind of abandonment. Even just asking, 'How much am I getting paid?' or 'When is the fee arriving?' can feel like a transgression. It's a minefield for nervous systems wired to fear punishment for stating simple needs. And so, compliance often feels safer than resistance – even when something feels deeply wrong. We may even lie to ourselves that we are being treated respectfully, in an attempt to push down the more painful truth. That's trauma, too: abandoning yourself to maintain connection. Thankfully, that's not Lisa's story. Her film, by design, actively resists the very hierarchies I've spent years growing wary of. Rather than being folded into someone else's vision, she was made a co-director – alongside Rebecca Lloyd-Evans, editor Alex Fry and co-writer Josie Cole. They were all credited equally, all paid the same. That flattening of the traditional film structure wasn't just symbolic; it was a genuine redistribution of resources and agency. It gave

Lisa space not just to tell her story, but to shape how it was told. That's no small thing. I read about it later, in an article describing *Blue Bag Life* as 'revolutionary' in its production model. I don't use that word lightly, but something about the way Lisa and her team went about this work – collaboratively, transparently, without the usual power dynamics – warrants the term. It showed what's possible when lived realities aren't just mined, but valued and respected. When storytelling isn't extractive, but mutual. Still, I can't help but notice how slippery this empowerment can be. And I can't help wondering what Lisa's career prospects might look like, should she pitch another film or project that doesn't centre or draw overtly from her own suffering. Her collaborators seem genuine in their desire to put money where their mouths are, but the example they set is rare. I also know where I sit. A little deeper behind enemy lines. Suspicious. Protective. Maybe even a little bitter. It's hard to know if my distrust is warranted or stems from some lingering unhealed part of me. Watching the crowd, I feel like a brother dropping off his sister at a prom – hopeful for her, but mistrustful of the room. Perhaps that's uncharitable. Perhaps it's projection. I can't speak for every storyteller, least of all Lisa, but that sudden empowering surge of personal agency remains elusive, save for the brief taste you get occasionally when something comes off well. Lisa is likely experiencing her brief surge right now, each eye in the auditorium fixed upon her, ears straining to catch her every word as I sit here in judgement of the host and much of the crowd, whether right or wrong. Some part of me cannot completely give over to what appears to be a room full of nothing but goodwill. Perhaps some will accuse me of being ungrateful, but the truth is, in Lisa's position, I occasionally feel like an animal caged in a zoo; aware I could overpower my handlers and flee if opportunity presented itself, though what I'd do, or where I'd go after, are unknowns yet too frightening to face.

*

Let's take a little trip back in time to 2018, the year my story-telling star rose higher than any other. It's a winter Aberdeen evening and the pharmacy is quiet. When closer to home, I am always careful in selecting a branch, my choice based usually on the number of weeks since the last visit. At any one time, I'll have three or four chemists on rotation to avoid suspicion. At local branches, such as the one around the corner from my home, or the three in the shopping centre in town, or the one just off the roundabout behind the shopping centre, or the superstore at the retail park, you're likelier to rouse concern (and the awkward questions that come with it) because staff are more familiar with their regular clientele. It is therefore useful, when attempting to acquire my current drug of choice, to supplement visits to local chemists with occasional stops at branches in city centres, where maintaining anonymity is less of a challenge. Thankfully, on this morning in 2018, given that I'm 145 miles north of Glasgow in the city of Aberdeen, the pitstop to restock before work is stress-free. Novel even. Regardless of whichever chemist I chance upon, wherever in the UK I happen to be, my routine is the same: upon entry, I move swiftly to the men's hygiene section (though not so swiftly as to draw the interest of security) where I pick up some deodorant and mouthwash, perhaps inspecting a fashionable razor I have no intention of buying, thus creating the illusion I am a normal consumer. Not that anyone cares. Then again,

I have been recognised in a pharmacy before – by the pharmacist. 'Are you Darren McGarvey?' she asked politely. 'Most of the time,' I replied before quipping that I prefer chemists indicate that they know me before and not after I purchase laxatives. Though having escaped the quicksand of social deprivation by recanting tales of trauma, stitched together with pointed social observations and my best attempts at political commentary, I cannot escape the sense that I feel the way an artist might do if they awakened one morning to find themselves trapped in their own drawing. Just a few months back, a 30-minute television special in which I was featured positively hailed me as a shining example of someone who 'overcame' their difficult start in life. Someone who 'transcended' their problems and whose story provides endless 'inspiration' to others still languishing in poverty's jaws. As I sat on my couch that autumn evening, cringing at the sight of myself onscreen while the handful of pills I swallowed 20 minutes prior began taking their increasingly limited effect, rather than a sense of achievement, I felt like a contemptible fraud. The queue today is tolerable, which makes joining the end of it easier. I move to the counter with a rehearsed expression, slightly more relaxed than usual as there is no one behind me – this transaction is easier to complete in the absence of prying eyes. I approach the pharmacist and ask for the painkillers. In this heightened state of awareness, mistaken assumptions about what another person may be thinking come fast and easy. I put them out of mind like I did the voice only seconds before, careful not to exhibit through body language what's going on in my mind. She turns to the display cabinet behind her, where the tablets are kept. I become slightly agitated as she walks the short distance to the shelves behind her. The shelves where they keep the good stuff. She selects a box of 32 soluble Solpadeine Max and places it on the counter in front of me before launching into the scripted spiel about the 'potential risks' that come with this medication. 'Is this medication for anyone who is pregnant?' 'No.' 'Are you on any other medications?' 'No.' She rambles on for what feels like two minutes, maintaining eye contact throughout. It is an unpleasant thing to experience; lying to someone's face is always easier when they don't look directly at you. Fortunately, I have experience in this area, too. As she enters the final furlong of a preamble that she seems even more fed up of saying than I am of hearing, I wonder how many times a day she is forced to partake in this strange little dance. For she must know – as I do – that as many people buy this product because they are addicted as do for the short-term management of pain. I swipe the box from the counter, tap my card on the reader and the tablets vanish into my trusty backpack – my closed-mouth friend in situations like this. If only these painkillers were as easy to give up as they are to purchase. On my next stop, I will pick up the ibuprofen equivalent. By then the laxatives will have kicked in and I'll be able use the bathroom where I can check if I am bleeding internally. If not, I'll take more pills, then nip around the corner to join my unsuspecting colleagues before setting off to continue work on a BBC documentary about poverty featuring me, Darren McGarvey – a lived experience success story.

Back in the present, the school run isn't even the hardest part of my day. If I'm lucky, the emails won't begin filtering through until 10am. Whatever my workload looks like, whether a public appearance, filming for television, recording music, or, in the case of today, working on a book, part of the labour involved for me is the unwritten rule that whatever

I produce must contain an element of autobiography. If it's the titles sequence of a documentary film, for example, then simply looking down the lens and telling the viewer what the films are about won't cut it. At some point throughout the 30-second series of clips, where I offer a taste of what's to come, a moment of personal testimony must follow: I grew up in poverty; I've experienced homelessness; I am an addict. Not only that, but those admissions must first be captured on camera, often outdoors in public places, with other people around – the language of television demands this. 'My name is Darren McGarvey, and I am an addict.' An old couple shuffles past me, glancing back on George Street, Edinburgh city centre. 'My name is Darren McGarvey, and I am an addict.' A group of students stop and stare, University Avenue, Glasgow's West End. 'My name is Darren McGarvey, and I am an addict.' 'I am an addict.' 'I am an addict.' 'I am an addict.' Then there was the time a journalist quoted the price I paid for rehab – one of a slew of indelicate moments where the objective seemed to be to probe me like a politician, perhaps with the aim of getting an unsettled reaction. Questions like, 'Why didn't *you* end up on heroin?' – my mum was an intravenous drug-user, you see – and attempts to draw me on a previous account of abuse, which I'd already made clear I did not wish to discuss. Most of the time, I'm fine with this slap-and-tickle, insensitive as it is, because I understand why I'm doing it; I'm not being manipulated but making a compromise with a firm under-standing of the terrain and my role within it. But sometimes, I wonder if the language of media is less about the hard rules of sophisticated art forms and more about a cheap, easy-to-throw-together recipe, which places the needs of a broadcaster or production company above the human beings involved. As

if the average punter just isn't capable of accepting that someone with a regional accent might have their own series simply
because they are moderately talented and work very hard. The
only way it could possibly make sense that I am presenting a
series about addiction is because I am an addict. Sometimes
I look at my trophy cabinet of writing and broadcasting
awards (I don't have a cabinet – they gather dust around the
house), my honorary doctorates (I know these do not make
me a real doctor or an academic and I don't keep them, I give
them to my dad). I reflect on what I've achieved as a qualified community arts practitioner who worked on the front
line for years, a trained journalist whose work has had a real
political and cultural impact, and a seasoned campaigner with
thousands of hours of grassroots work under my belt, and
consider the genuine effect on people's lives my work has, and
I think to myself: *Will I ever just be accepted as a professional
in my own right, like everyone else?* Why must I always be
framed as someone with lived experience, when I am often as
accomplished, across the various disciplines in which I work,
as anybody else in the green room? Do the people I work with
genuinely see me as a peer, or am I regarded as some interloper
or diversity hire, who hasn't earned his stripes like they have?
You could say this is just the imposter syndrome most people
deal with in life, but my paranoia is not entirely baseless. Any
time I've ever floated the idea of producing a piece of work
that was not about my poverty, addiction or some adjacent
topic, it's been met with resistance or disinterest. What this
teaches me is that my platform is conditional and flows primarily from my willingness to divulge details of my own life,
what I've been through and who I used to be – not where I am
now and the person I hope to become. Sometimes I return to

that Kevin Hines video, or to subsequent appearances he has made, and while I always marvel at his inspiring story, and see the value in him telling it, part of me feels like I'm looking at a man trapped in his own painting. A man who knows the game might be up if he should dare use his platform to tell a story about anything other than his suicide attempt. And in his sad eyes, I see a bit of myself. Like stars typecast in big-budget films that pay the bills, who long to play new characters that stretch them, but dare not step outside their sandbox in case they're relegated to the straight-to-DVD market. I have been blessed with wonderful collaborators across every profession I've been fortunate enough to get an opportunity in. Editors, directors and publishers who've held real space (even within the constraints placed on them by their respective industry orthodoxies) for my vulnerabilities and my thoughts, as well as my creative input. The problems I experience are rarely about the individuals I'm working with at any given time – they arise from long-standing cultures within the industries themselves – and my collaborators often deal with their own challenges arising from these very orthodoxies, too. But the personal disclosure aspect of my role is one few of my collaborators have any real experience with. You risk being seen as ungrateful if you complain too much, or lacking self-awareness if you portray yourself as some kind of helpless victim. This life is the bed I made as a storyteller, and I fully accept that – but by God they sure love making you lie in it.

Many people with a public profile eventually require the service of an agent to create a perimeter around them. This is due to the high level of correspondence public figures receive, and how difficult that becomes to manage. I'm far from a household name, but the volume of people and organisations that try to reach me, personally, over the course of a year would startle even the most seasoned celebrity. People contact me in desperation, begging for advice or help. Others make general enquiries with no specific request or pitch, just a stated desire to get on the phone or in a room to see what we could do together. When I am out in the world at community events and what not, I usually arrive not long before I am due to speak and leave quickly once I'm finished. People occasionally want more than I have time or energy to give. Often, agreeing to do something means agreeing to be asked to do additional things on top of that. And when you're out and about, you're accessible to others who may suddenly become struck by a desire to fold you into their plans. Point being: if I can limit my time out in the world, I reduce the chances of being flattered into agreeing to things in the moment which I may later regret. It's amazing how rapidly your schedule can fill up with 'stuff' and how much of that stuff involves me giving a lot and receiving little if anything in return. My main reason for having an agent is simply that I need someone to protect me from my own people-pleasing instincts. My agent is not like me, you see. She doesn't care what other people think. She doesn't require the validation of strangers to motivate her or boost her esteem. The day we met in London, the morning of my Orwell Prize win in 2018, she arrived wearing a necklace with the words 'FUCK OFF' written on it. Many people might have found that off-putting, but I knew straight away she was exactly the ally I needed. Beneath the angry, self-righteous and combative media image, I am in truth deeply conflict averse and often unwilling or unable to assert my needs or desires for fear of putting other people out or being thought of as difficult. In order to navigate the world I now inhabit, my health and wellbeing depend on the sound advice of at least one person I can trust completely. Someone who will tell me when to say 'no'. Someone who knows me personally and understands my vulnerability in the face of chancers, users and charlatans I often struggle to discern because I naively assume the best in everyone – or push down nagging doubts to maintain connection or gain affection. She is someone I can safely assume will act in my interests because those are interests that she shares. Her job is not simply to negotiate on my behalf; she takes a specific no-nonsense tone which establishes from the outset the kind of healthy boundary I often struggle to draw. Her approach sets the tone in her office, meaning whoever is working on my behalf on any given day understands the particular variety of interest shown in me and the myriad requests made for my time. An agent is basically a burly bouncer standing by your diary, drawing people dirty looks, occasionally frisking them. Without that perimeter, I'd become quickly overwhelmed. I want to be liked. I want to comply. I want to appear humble and never too much trouble and am therefore extremely vulnerable to exploitation, manipulation and psychological abuse – three specialties baked into the business practices of most media outlets in Britain irrespective of the many decent professionals working within them.

With all this work, most people only glimpse the finished product. That's partly why this gig appears from a distance to be such a blast. I get paid to go on and on and on about topics I care about and, of course, (my favourite subject apparently) myself. Behind the scenes though, this gets really tough, and the challenges are compounded by the fact I feel like I'm out here on my own most of the time. I don't move around in a pack like many of my campaigning contemporaries. I'm not an industry brown-noser you'll find holding court at networking events or awards shows. I'm the solitary figure usually smoking outside by a bin. Even people in this game with no major history of adversity or trauma will struggle at points. And for people like me, who carry with them a level of vulnerability, this is doubly so. While I must also make a living like the rest of you, I don't do any of these things for clout or money. Yes, there are lucrative opportunities at times, but please believe me when I say: if I was the grifter some online blowhards seem to think, I'd be worth a lot more than I am, I assure you. I enjoy what I do, and I acknowledge the privileges I now enjoy, but this is also extremely precarious and so very tiring for a nervous system wired like mine. You have to deliver the goods every single time, or at least that's how it feels. There exists pressure from every angle and at every level, whether debating politicians in front of an audience of millions or working from home on the next project. Even with agents or publicists dealing with a lot of things on my behalf, they can't hold my hand through everything. I still have to make decisions which always feel high stakes. The reason for this is pretty simple: every single thing I film, write or record eventually goes public and it will contain an element of personal disclosure. Can you imagine the level of vanity, self-loathing and fear that goes along with that? At time of writing, for example, I am preparing for a UK-wide press campaign to publicise my latest film series about inequality which will likely see me appear in a slew of magazines and newspapers, as well as on political talk shows and panel discussions. Then, after the shows go out, they will be reviewed and discussed not simply by professional critics, but by thousands of people online. Even now, as I write this, my heart rate is increasing as I begin to anticipate and try correcting for everything I believe could go wrong. Despite whispers that I am some kind of media whore who does everything for 'attention', I decline most media requests I receive; I've also declined opportunities to sensationalise or cash-in on my story in any manner I consider tasteless or vulgar, feeling its resonance thus far is based in part on my retaining some basic integrity. I know the less sympathetic among you will be wondering why the hell I don't just stop. I ask myself this question, too. But this isn't a life you can just walk away from; withdrawal would have to occur in phases, or I'd plunge my family into a financial black hole. Imploring me just to walk away would be a bit like me smugly telling you just to downsize to a smaller house or car during a recession – it makes sense technically, but isn't very practical advice.

As someone who has achieved a lot, sometimes I forget how much work I still have to do. When I tell people I am recovered, all I mean is I don't drink or take drugs, not that I am the finished article. I am grateful for the life I've been given but it's not a life I would necessarily have chosen had I understood more about my own woundedness, and what drove me unwittingly to turn myself into a product by simply telling my story. While some may feel I am complaining, I outline my experiences here simply to illustrate what really goes on when the curtains close and the cameras are switched off. I am trying to paint a truer picture of what high-profile lived experience campaigning truly entails. We often place ourselves in the very positions that we later come to find overwhelming. We must bear responsibility for those choices. That said, it must also be understood that there are immense emotional and psychological tolls behind many of the stories you encounter. My story is no different.

'Being a public person who talks about
pain is like putting your wounds on
a stage and hoping nobody applauds.'

Nadia Bolz-Weber, *Shameless* (2019)

A Story Going Nowhere

*What happens when the
story we tell ourselves no longer
aligns with reality?*

The day finally came when all my stories, narratives and false identities came crashing down under the weight of their own contradictions and my long-suffering sister dropped me off at rehab. She collected me from a police station on the west coast of Scotland where I was being held after being found at a train station, swigging from what was left of my bottle, staring a little too hard at the tracks. Having booked my bed in mid-October – at that time struggling only with codeine and occasional, small amounts of alcohol – I had very much expected to arrive in relatively good health, to undergo a brief clinical detox. But in the intervening weeks, my condition deteriorated rapidly, culminating in a three-day alcohol and drug binge which drove me to the brink of suicide. I'd been living a lie. A lie that I came to believe. And the truth always finds a way. Hardly known for my discretion, I announced my intention to take my own life on social media to tens of thousands of followers, rather insensitively invoking the late Frightened

Rabbit front-man Scott Hutchison's final tweet before he ended his: 'OK, I'm away now.' I don't remember composing the tweet, though I do recall feeling certain it was over for me. Scrolling anxiously through my inbox a few days into my stay in treatment, excruciated by what little remnants of my online behaviour I could bear to look at, I could also see that I'd replied to the countless messages of concern and distress that flooded in as my mental decline became more evident, with five chilling words: 'I wish I was dead' – a phrase that had become a mainstay of the negative self-talk over a two-year period in which my life appeared to change beyond all recognition. On the surface, I was doing well. My first book had become a bestseller. Film and TV companies were darkening my door to try and acquire the rights to turn it into a gritty drama. I got my own documentary series and wrote and performed in my first one-man show in a full run at the Edinburgh Fringe. Essentially, I found myself in the rather bizarre situation where every job I had ever fancied was handed to me on a plate within a 12-month period. Big cheques were arriving every other week. I was a 'success'. But somehow, this little phrase, 'I wish I was dead', still burrowed its way into my subconscious when my back was turned. There were even times when I caught myself saying it out loud, unaware of its true resonance. Trapped in my little tornado, and isolated from my old life, no longer able to relate to the people who truly knew me, and feeling they could no longer relate to me, I became a ticking time bomb. I told myself for a while that it was just a slip. That everything was under control. In truth, I hadn't been able to get any more than 90-days clean since November 2017, which marked the publication of the first book, and almost three years since my last drink. After two years of writing it, while learning how to raise a child, working multiple jobs as a freelancer as well as moving home, November 2017 was when I was going to try and take a break. How naive I was. All I had to do before stopping for a rest was travel to London for two media appearances. It's hard to describe what 2017 was like, prior to that London trip, but with the benefit of hindsight I now feel comfortable referring to everything before it as 'my previous life'. The London trip was when everything changed. I had two main engagements on my trip, the first an appearance on Radio 4's *Start the Week*, on the Monday morning. The second, an appearance on Jeremy Vine's show on Radio 2 the following day. Given that I had been so busy dealing with the publication and launch of the book prior to the London trip, I hadn't given these radio appearances much thought beyond the fact they were happening. It may surprise some of you to learn, for example, that I had never listened to *Start the Week* and first heard of it when the booking to appear came in. I had certainly heard of Jeremy Vine but was not aware of how many people listened to his afternoon show. In a way, this worked in my favour, as on the Monday when I arrived at Broadcasting House, I was extremely relaxed. It also helped that I was relatively unknown anywhere south of Gretna Green, so I did not feel any great weight of expectation. When it came time to take part in the live broadcast, which featured three other guests, all I was focused on was giving as good an account of myself as possible. One guest was a composer, the other a theatre director and the third an epigeneticist. It struck me as odd that I had been invited, given how successful, experienced and intelligent the other panellists were. *I'm nobody*, I thought to myself, *why am I even here?* Of course, by then I had years of practise at concealing that sense of being less than. In Scotland, the book was flying off the shelves thanks to strong word of mouth. That would have been enough for me. The London visit seemed to me a long-shot. I was already exceedingly aware of how little genuine interest London media has in regional matters of politics, and even less in the arts, never mind the small issue of my sounding Glaswegian and working class – two traits that work against you in most places, but especially among the London media class. Still, I figured the publicists who secured the slot must know a thing or two and so, in an act of deference to their superior knowledge, I decided just to throw myself in.

Throughout the course of the 45-minute conversation – of which I was allotted around 12 minutes – I held my own pretty well considering I was, by every measurement, the least quali-fied person in the room. It went so well in fact that when other guests were prompted to discuss their own work, they would often direct the conversation back to me, which provided me with additional opportunities to expand on my ideas. I then realised they had all read the book. Weird. The 45 minutes flew by, feeling more like 20, and when we came off air, I felt a real sense that I had risen to the occasion. At no point, either before arriving in London or prior to going on air, had I even slightly considered that *Start the Week* had an audience of millions. That it was one of the biggest shows on radio. At no point had it occurred to me that its audience was used to hearing successful, middle-class professionals expound upon their lives and work, and that my voice, accent, use of language and general insight, drawn mainly from life experiences, would stand out so much. But that is exactly what happened. And in that moment, all of my handicaps became assets. Everything that had previously held me back would soon propel me onto an international stage. Only when I left the BBC, bungled into a cab to head to a nearby bookshop, did I begin to get a sense that my life was about to change in quite a dramatic fashion. In the hour or so that I had been in the studio with my phone switched off, thousands of notifications had been pouring in on social media. Thousands of notifications from people all over the UK, who had just heard me on the radio. In the back of the taxi, I thumbed through the likes, comments and messages, but every time I thought I had got to the end, the screen would refresh and more would appear. I remember thinking my phone was broken or that Twitter was malfunctioning. Emails started pouring into my inbox and junk-mail folders. Friend requests, follows and messages flooded my social media. My phone seemed to be beeping, vibrating or ringing almost constantly, alive with the activity taking place online. For many people, this would be tremendously exciting, to know that so many people were interested in them or their work, but for me, it triggered an unbearable state of mania; my heart raced, and I became mentally and physically disorientated, the world suddenly seeming unreal, as I located myself in that moment many people dream about: the moment when you 'make it'. My heart was racing by the time we arrived at the bookshop. Soon after, the publicist left and I was, for the first time in a while, all on my own. I ordered some food and sat down to catch my breath, as the notifications continued to swamp me. It was then that I received an email from my publisher, Gavin, containing a screenshot which, at first glance, appeared like it must have been doctored in some way for the purposes of a joke. The book, if this screenshot was real, had climbed to number nine in biographies on the Amazon sales chart. In a state of complete shock, I opened my laptop and searched for the book online, desperate to see it with my own eyes, but by the time I located it, the book had risen to number one in biographies and into the top ten for all books. When I say 'all books', I mean every single book available to buy on Amazon in Britain. How was this happening? Struggling to see, I put my glasses on frantically to double-check I wasn't misreading it; to my astonishment, as indicated by a special logo next to the title, my book was a '#1 Best Seller'. Even now, as I write this, the emotional memory is vivid, but rather than pleasant, the feeling registers more like a flashback to a traumatic event. I sat with my head in my hands, almost unable to bear the reality of what was occurring. This was more than I had ever dared to think possible. I looked around me, and everyone was just getting on with their day, blissfully unaware of either me or my stupid book. But online, it was as if someone had rammed a foot down on the accelerator pedal of my life before I'd had a chance to fasten the seatbelt. Again, I looked around me, overcome by a natural urge to share the news, but I became acutely aware of the fact that I was alone. In truth, I had grown increasingly isolated throughout the challenging process of completing the book. And the more consumed by it I became, the less able, or willing, I was to talk about my difficulties. It's such a rare experience to go through, especially where I come from. No matter how deeply people around you care, you get the sense that something fundamental is lost in translation when you talk about writing a book – unless you're talking to another author. Back then, I didn't know many people who had written books. And any acquaintances I did have, while supportive in ways that I'll never forget and for which I'll always be grateful, were not familiar enough that going to them with my personal problems seemed appropriate.

So, you suffer in silence. It becomes easier to say as little as possible. Then there's the additional problem of not wanting to sound like you are complaining; getting the opportunity to write a book, never mind for it to be published, is such an immense privilege.

I became gradually more isolated from family and friends, despite them always being close by and extremely support-ive. I began feeling like I couldn't be open about how I was struggling with self-doubt, exhaustion and fear. My problem, I reasoned, was so petty and indulgent compared to their adversities. What catalyses the downward spiral is the belief that your experience is different or exceptional in some way and that your problem can be triaged, for treatment later because there are far more pressing and important things than your feelings to consider. That sense, that others are unlikely to understand, is what creates distance between you and the people who are best placed to support you – the people who know and love you. By the time the book came out, I was mentally and physically bereft, running on pure adrenaline and caffeine. The strain I was under prior to its release wasn't just because of the immense seven-day-a-week workload, or the sleepless nights and early mornings. It also came from the pressure I'd put on myself to produce a good book. The weight of expectation its imminent publication had generated, especially among those who had been following my work for years before the media took interest, and those who kindly supported it through a crowd-funder. But perhaps the most acute pressure was that I now had a family to raise and provide for. I became obsessed with the idea that my son had to have a certain quality of life. That he had to be insulated entirely from the social and economic adversities that had shaped my childhood. The childhood I drew on in my writing. Sitting in the bookshop, suddenly this pipe dream of breaking the cycle of poverty and adversity began to seem possible. I could taste it. By that evening, thanks to *Start the Week*, I was booked up with enough public speaking gigs, festivals and appearances to last 18 months, but my expectations – realistically low in the morning – had now shot through the roof. It's not hard to see why of course – I had sold a few thousand books in a matter of hours. I went from not having any expectations of what the trip meant, to sensing that I was on the cusp of a profound and life-altering moment. A window into a new quality of life opened before me. A life where I may be able to provide things for my children that I never had. A life where I might become the person on whom others felt they could reliably depend. A life that was not defined by poverty and adversity but one where I might begin to derive a sense of value, self-esteem, and emotional security from doing what I love.

All that was left to do was Jeremy Vine's show the following day. That became the centre of my focus, as I marshalled my remaining wit and energy, heading back to my hotel, across the road from Broadcasting House, to prepare for the next morning. But on Tuesday, just over an hour before I was due to be interviewed, I received a phone call informing me that the appearance had been cancelled. I was broken by the news. It was soul destroying. Imagine dribbling the ball from your keeper's goal box, all the way up the field, past every player on the opposing team, only for the referee to blow the final whistle before you could take your shot on an open goal. It was crushing and felt extremely personal. Whoever pulled the plug on my appearance wouldn't have had any idea of how vulnerable I really was, but not even I knew that. How was anyone to know? Nonetheless, I felt humiliated, having been trailing the appearance all week and the night before to the increasing numbers showing interest in what was going on. Looking back, I can see that I was off my head. The idea that my future rested on one interview at the BBC was always silly, but when you are at the centre of a self-generated tornado, perspective isn't something you can rely on. I was a failure. The dream was dead before it had even begun. And then that bubble of manic energy in which I had become cocooned suddenly burst, sending me hurtling back to reality. I don't recall receiving an explanation or a reason. No attempt to rearrange was made. This I now understand is pretty standard but, back then, I was inexperienced and found the impersonal nature of it all extremely hard to accept. And again, as I experienced this intense flood of emotion, like I had the day before, I found myself completely alone. I had built this fantasy up in my head, believing my career, my children's future, the love of my wife and the respect of my peers were all conditional; contingent on becoming 'successful' – that's the story I was telling myself. When your sense of perspective has become so hopelessly skewed, and your spiritual defences so fragile, then, as an addict, it was not a matter of 'if' I would relapse, but when. The next thing I remember is being in a Boots pharmacy in Euston station, where I purchased a box of over-the-counter painkillers containing codeine, then frantically trying to find a quiet place where I could take them – an ambitious aim in central London. Moments later, sat in a nearby café, I was overcome by a sense that everything was OK. I felt a pleasant light-headedness, chased by a dash of euphoria, heightened by the sudden release of tension in my neck, shoulder and face. I had been walking around clenched for about 18 months, in a perpetual state of tiredness, fear and dread. Now, everything wasn't so bad. Four tablets wasn't a big deal, surely? I'll throw the rest of the box in the bin before I board the train and act like nothing happened. Of course, the codeine high wore off quickly and the precise second I felt it fade, I looked at my watch to work out how soon I could take more. Of course, the whole lot was done within days and by the week's end, I was taking both the paracetamol and ibuprofen together, to increase the codeine dosage. My bag was littered with torn packets, empty boxes and discarded instruction pamphlets. My digestive system was now in meltdown, meaning I was not eating correctly or going to the bathroom, sometimes for days. I decided the best way to deal with this was to resume smoking cigarettes, two years after quitting, and to use laxatives – a barrel of laughs when you're travelling. I smoked through the nicotine-induced headaches and nausea while shitting through the chronic constipation. In just a month, I dropped a considerable amount of weight but somehow couldn't see I was in trouble. I couldn't see I was in danger. The notion that I was experiencing a mental health crisis or in the grip of a relapse into substance misuse faded into the background behind the more urgent concern: the immediate relief of the emotional and physical discomfort brought on my drug addiction. All of this went on for months, as I appeared at trauma-informed events, espousing the principles of recovery and healing. By the time I reached rehab, after over a decade of experience in recovery, that man felt like another person. Somebody I had to begrudgingly take responsibility for. As the various substances began leaving my system in the first two days, they were slowly replaced by the inevitable shame, remorse and fear which, in any other circumstances, would require more inebriation for a serious addict like me to bear. Thankfully, I remained in treatment for 28 days. Not even Librium – a benzodiazepine which helps prevent seizures – was enough to slow my disordered, racing brain. The torrent of self-centred despair that temperate drinkers might know as a hangover would have been a welcome reprieve. I thought I had seen my rock bottom. I thought the nightmare had ended years earlier, when I first got clean. Apparently, you can plunge greater depths each time you buy into that dangerous idea that you can 'just have one'. That 'this time it will be different'. Or, in my case, 'fuck it, I'm willing to pay the cost' – stories I told myself, all of them lies. For all the pain, it wouldn't hurt as much if I were the only person affected. But this is rarely the case where alcoholism and addiction are concerned. This condition burrows its cunning little tentacles into everyone around you. My children may have been too young to recall this episode, but I was under no illusions.

One support worker, during one of my early therapy sessions, shared his own experience, quipping, 'If you shook my family tree, bottles would fall out'. I laughed, but only to conceal a realisation that I, the son of an alcoholic mother, who was the daughter of an alcoholic mother, was by then treading very dangerous waters as a parent. At various points throughout the previous months, I had contemplated suicide, but it began more as a playful idea. A movie spooling in my mind. Not something I took seriously. A bit like the idea that one day I might have to lift a drink if someone I cared about died suddenly. But these stories, however fantastical, take root if contemplated too obsessively until your entire field of vision is dominated by nothing but delusions accompanied by increasingly narrow and terrible options. The day then came when someone I cared about did die – by suicide. Before I knew it, I was guzzling painkillers, sure I would stop the next day. A month and a half later, having transgressed so many of my own moral lines, whether stealing drugs or being dishonest, I was walking around the streets of Glasgow crying, my neurocircuitry misfiring, breaking down, as my mind and body coursed with so many different chemicals that I slipped in and out of even knowing who I really was, while sipping from a Lucozade bottle filled with Buckfast Tonic Wine – psyching myself up to leap the railing of a bridge, or bound in front of an express train. But at the same time, I was also intermittently aware that I was very unwell. That my thoughts were not my own. At various points where I could feel some build up to an explosive act, I became suddenly able to observe my sickness, while simultaneously being in its death grip.

Unlike many who don't make it through such a crisis, I grew up in a time when a lot of help was available: child psychologists who taught me how to disidentify with my thoughts, cognitive behavioural therapists who told me that I was not the story I kept retelling myself, and youth workers and mentors, who spent many days and hours and months with me as a youth, reaffirming what was wonderful about me and why I had reason to be hopeful about the future. A reason to stick around and finish my story. The kind of support you don't get anymore. The kind of answers most people go looking for online, at some ungodly hour of the day. Today, it's only when you are in a genuine crisis that the systems around you become suddenly active. Thankfully for me, this mental struggle raged for just long enough that help arrived, as it often has for me, in the form of a police car. I believe I came undone because I had become untethered from reality. I was viewing my life through a lens chipped and scratched by trauma. With an intermittent sense of my own worth, a deeply unstable sense of identity, and a deep fear of being judged, making a mistake or letting people down, I gave everything I had to a pipe dream that if only I 'made it', I'd feel better, and everyone would accept and love me.

In Part 3, we will separate out the various strands of falsehood which came to dominate my inner world, placing my nervous system on high alert, and the lessons I learned during my stay in rehab – where my delusions were once again smashed. What must be understood, however, is that I walked around for two years in great emotional distress, and nobody was any the wiser thanks to my mastery of this public performance. Despite being surrounded constantly by professionals with deep insight into trauma and mental health, few ever acted on the impulse – which I am sure some of them felt – to ask me if I was *really* OK. Nobody ever challenged any aspect of my story. I was affirmed, validated and applauded until I literally could not function under the weight of my own bullshit. I don't blame anyone, of course. After all, the culture around trauma as a matter of principle always believes the version of events as told by the victim. But affirming someone's story is only the beginning of truly supporting their healing. Further along that path, there may also come a time for harder truths. It's those harder truths that we now approach as we begin the final part of our journey. When the aim is to support a person to recover, and there is a possibility that part of the victim's narrative may be incomplete, or even flat-out inaccurate, in ways that impede recovery, as so many plot-threads of my story have been, does it not fall on the movement that helps itself so generously to our stories of personal adversity to develop some mechanism whereby the delicate matter of the actual truth (and not a sanctified personal truth) can be delicately broached? And in the event that we crash and burn as I did, who is ultimately responsible for clearing up the wreckage? While the Trauma Industrial Complex is relentless in providing context, reason and even justification for our personal calamities – some argue they are rooted in systemic inequalities while others promote self-help and introspection – the notion that it was ever anyone's responsibility but my own that I ended up in that desperate mess was arguably the most dangerous delusion of all.

PART 3

Releasing

'Trauma shapes how we see the world, not because it's accurate, but because it was once necessary for our survival.'

Gabor Maté,
The Myth of Normal (2022)

CHAPTER 10

Turning the Page

*Where do we begin when
rewriting our story?*

By now, if I've done my job, facilitators, consumers and storytellers alike should feel a bit uncomfortable about their entanglement with the Trauma Industrial Complex. Many of your sacred cows lay slain in my abattoir: not every feeling we have is valid, not every concern we express is authentic, and not every story we tell ourselves or others is true. Often, the stories we cling to are just another way to avoid the hard work of healing – fantasies we retreat into where everything makes sense, and our versions of ourselves and our lives are beyond reproach. You've spent the last few chapters in a narrative trance I deliberately induced. Now it's time to gently awaken you from your hypnosis and remind you that you are in fact reading a book. A book written by a storyteller who knows this terrain intimately. Someone who understands the craft of storytelling, is humble enough to admit he doesn't have all the answers, yet grandiose enough to refer to himself in the third person. Not everyone who started this book will have made it this far. Many likely bowed out a few chapters ago, unable to sit with the discomfort of my perspective – let alone the notion that the parts of it that irritated or upset them might hit closer to home than they're ready to admit. That's fine, though, because this book was never meant to be for everyone. Just as there is no one-size-fits-all treatment for trauma or a stable, cohesive definition of the term, there are countless ways to navigate it. Healing – by which I mean learning to live and thrive as trauma walks alongside us – requires us to explore various options available before choosing one we can make an honest go of. This book should act as a companion whatever choice you make and is written for those who sense that the stories they've been telling themselves may be incomplete – and that now may be time to perform a page-one rewrite. Those stuck in incomplete narratives about trauma often approach the Trauma Industrial Complex like fad diets during a health kick. The buffet of options is so vast that the labels we assign ourselves can feel liberating at first but may quickly morph into dangerous new controlling ideas. These stories can temporarily quiet the unease of being unmoored, offering a fleeting sense of purpose or meaning. Often, we gravitate towards what feels good, and what feels good is usually what seems easiest. Someone with ADHD might pursue a mental health diagnosis to avoid grappling with the unsettling implications of neurodivergence. An alcoholic might seek a neurodivergence label to sidestep the more confronting reality of their drinking. Likewise, those who sense they carry trauma may shop around for a label or framework that aligns with their identity or narrative, offering relief from the need to engage in deeper self-examination. After all, who wants to admit that their trauma manifests as controlling behaviours in romantic relationships? Or that they struggle with emotional maturity as a parent or spouse? Or worse, that they've become manipulative or abusive? The term 'trauma' is overused today, but this is how it often surfaces – long after the original event. Trauma shows up in the very relationships we turn to for comfort, bending them to its will. In these moments, stories become convenient plasters for deeper wounds. It becomes easier to focus on what was done to us – on blame, unmet needs or whose fault it all was. While much of this might be true, the story rarely ends there. Unprocessed trauma often leaves us emotionally unsafe, even toxic, to others. Yet, in our heightened state of vigilance, it feels unthinkable to emerge from our trench and confront how our pain has mutated. That's the true meaning of 'hurt people hurt people' – a phrase we often use to describe others but rarely turn inward and apply to ourselves. Instead of doing this hard work, many people lap the Trauma Industrial Complex until they find a shiny new framework that feels empowering but ultimately serves as a superficial fix. These labels often come with complementary ones for the people in our lives who challenge us. It feels good for a while, but like a January fitness plan, the novelty wears off, and we begin to tire of the introspection. Much like sticking to a diet or exercise routine, healing from trauma requires awareness, consistency and effort. Most approaches work if we commit to them, focusing on understanding our patterns and implementing healthier ones. Progress is rarely linear, but it's the gradual gains – often following moments of real exertion – that encourage us to keep going. Trauma, however, whispers for us to stay in our comfort zones. It prefers us pliable to its suggestions, steering us away from the hard labour required to take control of our lives again. But healing demands we resume the driver's seat, charting a new course – one that may be uncomfortable but is ultimately liberating. If you're ready to embark on this path, know that it will not be easy. It will demand courage, commitment, and a willingness to move beyond the comforting narratives that have held you back. But in doing so, you may discover a freedom far greater than the illusion of safety trauma offers you – a freedom that is truly your own. Trauma healing often unfolds in three interconnected stages once safety has been established: awareness building, processing and releasing. These stages, while not strictly linear, represent a progression towards integrating traumatic experiences and reclaiming a sense of self. Influenced by pioneers like Judith Herman (*Trauma and Recovery*), Bessel van der Kolk (*The Body Keeps the Score*) and Peter Levine (*Waking the Tiger*), this model underscores the importance of addressing trauma holistically. While you'd struggle to find a take on trauma which is not in some way contested, it's widely understood that healing from it follows the trajectory outlined here, perhaps with some minor caveats or different language.

The awareness-building stage involves cultivating an understanding of trauma and its impact on the mind and body. Judith Herman describes this as the 'establishment of safety', where individuals learn to name and recognise their trauma symptoms. This phase often includes education about triggers, flashbacks and dissociation, enabling individuals to gain clarity about their experiences. Building awareness can also involve mindfulness or somatic practices, such as grounding techniques, which help individuals stay present and distinguish between past and present realities. This stage sets the foundation for deeper work by fostering stability and self-awareness. In the processing stage, individuals confront and begin to make sense of their traumatic memories. Peter Levine's somatic experiencing framework emphasises the importance of releasing trauma stored in the body through gentle, incremental exposure to distressing memories, rather than reliving them in full force. Similarly, therapies like EMDR and cognitive behavioural approaches enable individuals to reprocess traumatic events, reducing their emotional charge. Processing requires a delicate balance of accessing the pain without becoming overwhelmed by it. The goal here is not to erase the trauma but to reframe its narrative, integrating it into one's life story in a way that fosters meaning and growth. The final stage – releasing – involves letting go of trauma's hold on the individual. This may include releasing pent-up emotions like anger, fear or grief through creative outlets, movement or therapy. Practices such as yoga, expressive arts and bodywork are often cited as effective tools for helping the body complete the stress response cycle. By releasing trauma, individuals can cultivate a sense of empowerment and freedom, rediscovering aspects of their identity and life that were previously overshadowed by pain. Healing is rarely linear and often requires revisiting these stages multiple times. However, with each cycle, individuals can move closer to a life defined by resilience and authenticity rather than trauma.

In the context of this book, the three stages of recovery have also been loosely applied to the cultural conversation surrounding trauma itself – a conversation I call the Trauma Industrial Complex. In Part 1, we explored the rise of the lived experience, examining how storytelling has become a dominant force in shaping cultural narratives about trauma. We discussed the potential risks of making personal adversity public in a world set off its axis by digital revolution. We also delved into the role of catharsis culture and the excesses of an unregulated marketplace, where trauma-related content is commodified and consumed against the backdrop of a mass mental health crisis. In Part 2, having deconstructed the mechanics of the Trauma Industrial Complex, I used my own story to demonstrate storytelling's dual power – to elucidate and entertain. I subjected my own narrative to rigorous scrutiny, modelling how individuals like me – those with experiences of trauma, searching for answers but without access to formal treatments – might begin to untangle the 'little movies' playing in their heads. We also explored the potential unintended consequences for storytellers and those connected to them. Which brings us now to the third and final phase of our journey – releasing.

For me, much of the pain I carried (and still do) was wrapped up in stories: the story of my mum, the story of being working class, the story of trauma itself, and, eventually, the story of how I couldn't cope with the burdens of these stories so publicly. There was truth in all these narratives, but there was also a fair share of distortion – conscious or otherwise. Even when I began writing this book, I hadn't fully recognised the extent to which my narrative shaped, and sometimes limited, my understanding of that experience. I can only speak for myself here, but I suspect we may never arrive at a final, definitive draft of our personal story. Instead, we spend our lives rewriting it – sometimes improving it, sometimes complicating it. The best story, however, isn't necessarily the one that makes us feel good or resonates most deeply with an audience. The best story is the one that rings truest. In this context, releasing is about revisiting the finer details of the stories that have driven so many of our subplots, identifying the moments where narrative embellishment (or white lies by omission), brought on by old wounds have held us back. It's about finding the courage to let go of the parts of our story that no longer serve us and leaving them, as it were, on the cutting room floor. Releasing isn't about erasing the past or denying its impact. Nor is it about forcing ourselves to forgive and forget that which still pains us. It's simply about making space for new possibilities, and stories that are still waiting to be told – stories that move us closer to a more authentic, compassionate understanding of ourselves and the world around us. This is a process which may occur countless times. A painful but necessary process which began once again for me back in 2019, when I entered treatment for my alcohol and drug relapse.

*

I arrived at rehab a wreck – physically, mentally, emotionally – so dehydrated they couldn't draw blood from my veins. Exhausted, I just wanted to sleep. After a brief assessment, I was given medication for detox and taken to my room. For 24 hours, I was left alone to rest. Someone brought me toast and tea, which I devoured before drifting back to sleep. By the second day, the fog began to lift, and so did my self-pity. At the time, I didn't recognise it as such. I thought I was unique – a victim of my circumstances, misunderstood in my moderate fame, an experience complicated by my class background and trauma. Surely, no one else at the facility could grasp my specific, special predicament. Delusional. I had abandoned the recovery principles that had kept me sober for so long and my trauma at bay, allowing my ego to run rampant. I'd convinced myself my trauma was extraordinary, my situation singular. Thankfully, the staff weren't buying it. It was in the third week of treatment, during the tail end of a wearying group therapy session, that David, one of the two therapists working in the rehab, said something that hit me like a frosted pane of bullet-proof glass: 'It's not what happened, but the meaning that we make of it.' The session that day had been about trauma. By then, we had already navigated the widely recognised aspects of addiction – the craving for the pleasant effects of alcohol and drugs, the body's inability to process these substances without sparking an insatiable thirst for more, and the chaotic descent into lies, obsession and oblivion. We knew the cycle: the build-up of stress, emotions or adversity, the eventual 'fuck it' moment, and the catastrophic aftermath of relapse. But David wasn't focused on this predictable spiral. He was taking us deeper, urging us to explore the stories we told ourselves about who we were and why we drank or used. Six years into recovery, I'd grown used to having hard truths told to me, but this one stung. Recovery teaches you to surround yourself with people who care enough to challenge your fragile ego. I'd been called out plenty: for resenting my partner's lack of warmth while lying to her face, for judging others while masking my insecurities, and for painting my enemies in stark black and white while wrapping my own misdeeds in endless shades of context. It's painful to face truths like these, but you learn that your mental health depends on living in reality, however unflattering it may be. David's statement was a new kind of challenge. He wasn't just calling out my denial about alcohol; he was suggesting I'd also constructed elaborate narratives to explain my addiction, to protect myself from facing harder truths. It was a difficult pill to swallow. The years I'd spent cobbling together some semblance of sobriety had taught me to uncover my prejudices, own my manipulative behaviours, and recognise how I had unwittingly harmed others while claiming victimhood. Still, the fact that I had relapsed meant I either hadn't learned enough or had forgotten it all. As bleak as rehab felt, one reason gave me hope: half the staff were in recovery themselves. They were those humble lived experience people, going about their business quietly, changing the world a day at a time. Advice from someone who'd been through it carried weight. Without direct experience, you'll betray your ignorance – like someone without kids offering unsolicited parenting tips. So, despite my initial defensiveness, I trusted David and the other therapists to guide me. But it wasn't easy. 'No,' Marie, another therapist, said bluntly in front of a room full of fellow addicts. 'I asked how you feel, Darren. What you told me was a story.' Her directness was jarring. Marie saw through me instantly. She called out my tendency to frame everything as a narrative, a skill others had recently praised but that she saw as a defence mechanism. Her assessment was that I intellectualised my emotions as a way to avoid feeling them. This is also an observation made by partners at different points in my life – I express what I think are feelings, which are really just detailed observations or criticisms of how I believe other people behave. I don't focus on how my interpretation of their behaviour makes me feel, I focus on them and what I think they're doing wrong, and on what I reckon they need to do better. That might cut the mustard if you're a political activist, but in personal relationships – unless you're dealing with a Grandmaster in transcendental meditation – such an approach only gives rise to conflict. She identified my inner child, crying out to be seen and heard. That subconscious part of me that, when activated, spoke from a place of woundedness, perceiving the present through the lens of the past.

'You tell stories about your life to protect yourself. Why is that?' Marie's words infuriated me at first. I'd spent two years being invited to speak at length about myriad social and personal issues. My articulacy was often being remarked upon positively. My speech style in many ways had become as pivotal a part of my public image as my trauma or political opinions. But I'd grown out of practice with respect to the rhythm of real conversation; accustomed only to being given the floor. To simply be interrupted while enlarging upon my thoughts (delusions) felt itself like an injurious insult. I had been mollycoddled and pampered for so long, receiving praise, due and undue, from every direction. To suddenly feel the prick of Marie's pin, bursting my self-important little bubble mid-story, stung to say the least. She wasn't interested in my achievements or my public profile. Her approach was like alchemy. She would enter the room and improvise each session based on what was happening in the space. She could turn on a sixpence in the face of an unexpected emotion or revelation. And when I began my 'speech' about my life, she could see my attempt to make an audience out of my fellow addicts – a way of elevating myself above them so as to avoid the vulnerability of truly connecting. And she was right. I needed her honesty, even if it hurt. Over our sessions, Marie helped me get to the roots of much of my turmoil. I was deeply insecure in my relationships, particularly with my partner, whom I believed would betray or leave me, and due to slacking off on my recovery work, I had become ruled by fear. The fear I was about to be abandoned. The fear I wasn't good enough. The fear of public humiliation and shame. The fear of financial oblivion and the fear my children didn't love me. In this fearful mindset, every hypothetical situation was taken to an absurd extreme in my imagination and rather than attempt to gain perspective on it, I would begin mentally preparing for these worst-case scenarios in every area of my life. It wasn't until I experienced deep emotional relief at the thought of taking my own life that I began to sense vaguely something was wrong. I had practically talked myself into suicide, not because the circumstances of my life were materially difficult, but because the story I was telling myself was so unbearably horrific. It's no wonder I relapsed when you consider the pain one would experience when subjecting themselves to such an onslaught of self-hatred over a prolonged period of time. While any intervention must be mounted delicately, what a person with first-hand experience has in their armoury is intuition of when precisely to affirm and validate feelings, and when to probe or challenge them. Rehab is not an affirm-only environment. Affirmation is partly what got me drunk. Rehab is more of a 'your very life depends on the truth and not a story' kind of place. That's what some of us really need at some stage of this healing journey. Marie and David's good-cop, bad-cop routine stripped me of my delusions, day by day. They cared enough to dismantle my defences. I wasn't special. My situation wasn't unique. My problem wasn't alcohol or drugs; it was me. Alcohol and drugs were my attempt at a solution to that problem. The self-seriousness and learned helplessness had to go. Everyone suffers some form of trauma. Everyone carries some form of burden. Most people don't do what I chose to do and attempt to burn their lives to the ground while blaming everyone and everything but themselves. Thankfully, despite the serious nature of the work, Marie and David's sessions were a mix of humour and hard truths. 'Why do you keep taking drugs? You're really shit at it,'

David would say, laughing. Then he'd dig deeper. When I shared my childhood experiences, with a melodramatic drawing of myself as a jagged scribble, complete with torn paper, I expected sympathy or shock. What I got was akin to indifference. He refused to feed the drama, which in turn, began to break its power. Instead, he challenged me to rethink my interpretations. Bad things had happened to me, sure, but not always for the reasons I'd told myself. My mother didn't hate me, she was just too unwell to meet my needs as a child. By ascribing a malevolent intent to her behaviour, I created more pain and more unanswerable questions to torture myself with. Life was more complicated than my narrative about trauma, poverty and class allowed. Letting go of some of that narrative wasn't easy – I'm still struggling – but my eyes were at least opened to the notion that a new story might be possible. By then, though, the old one had become a cornerstone of my career. In industries that prized authenticity, my story had been my calling card. David's challenge – that my narrative wasn't the whole truth – felt like an existential threat. I could have reacted defensively, played the victim or stormed out. But I was desperate to get well. For my wife, for my kids, for myself. I was sick of hearing myself and sick of feeling anxious, depressed and resentful. So, I swallowed my pride and listened. I became willing to let the story go. We all have stories we tell ourselves about who we are and why. Events shape us, but it's our narratives – our interpretations – that give them meaning. Two people can experience the same circumstances and emerge with vastly different narratives. These narratives shape how we see ourselves, how we interact with others, and how we cope with life. They run like a news ticker in our minds, influencing every decision, every reaction. But what happens when the narrative is flawed? When it's built on misconceptions or self-deceptions?

Narrative therapy suggests that by examining these stories, we can rewrite them, creating healthier ways to understand ourselves and our experiences. As a therapeutic approach, narrative therapy encourages individuals to externalise their problems, treating them as separate from their identity. This process involves exploring the dominant stories that shape a person's life and uncovering alternative narratives that reflect resilience, strength and possibility. By reframing past events and challenging the interpretations that sustain negative self-concepts, narrative therapy can help individuals reclaim agency and rewrite their life stories in empowering ways. For me, this meant confronting the ways I had framed my life around trauma and poverty. My identity was built on a narrow, incomplete view of my story. Once that narrative began to collapse, I had nothing solid to stand on. The weight of my contradictions – a 'recovered' addict in relapse, an 'authentic' speaker lying to himself and everyone else – became unbearable. Marie and David's work was to break down my ego, to divest me of the falsehoods I clung to, and to make me once again teachable. It wasn't easy. The process is inherently uncomfortable, and you resist it without realising. But their honesty, rooted in their own experiences, made it possible for me to let go of the stories that were holding me back. David's statement – 'It's not what happened, but the meaning that we make of it' – was the most provocative thing anyone had said to me since my public rise. He essentially committed the cardinal sin of invalidating my story and trauma – and it worked. It forced me to rethink everything: my past, my identity, my choices. I had spent a lifetime constructing a narrative that framed me as a victim, a survivor, a man shaped by adversity. But I had also played a central role in creating the circumstances that overwhelmed me. I wasn't just a character in the story – I was its author. My narrative, when broken down into its constituent parts, contained four fundamental flaws I'd eventually have to reckon with. The first, that my story was true and mine alone to tell. The second, my desperate need to cling to aspects of my identity out of vanity, fear or judgement, and uncertainty of who or what I'd be in their absence – all while hiding other less savoury facets of my identity from view. The third, that I was a victim within this narrative rather than a collaborator in many of my difficulties. Finally, I became trapped in a delusion that fighting in a very public way for the world I believed was possible, was the same as healing. That simply asserting my beliefs and opinions, and holding the perpetrators of inequality accountable on social media platforms, was somehow a revolutionary act. In truth, much of my activism following the success of my first book was performed remotely, with very little graft involved – it was an image I wanted to project in the public sphere, in an attempt to relieve survivor's guilt. These are the four strands we will explore on the final part of our journey. Using my own narrative, I will deconstruct each flaw, holding a mirror up to your story in the process. Perhaps, together, we might find clarity and peace in some painful truths – even if they sting a little at first.

'I wanted to believe I could tell my story and be done with it – as if my story was mine alone. But every word was a rope thrown backward, tying me to something older, something shared.'

Ocean Vuong,
On Earth We're Briefly Gorgeous (2019)

CHAPTER II

Taking Ownership

*When we tell our story publicly,
does it truly remain ours?*

It was during an interview with *Channel 4 News*, the morning after my debut book won the Orwell Prize in 2018, that I began to worry I had bitten off more than I could chew. I sat with the programme's social affairs editor, Jackie Long, in the centre of large, dark, wood-finished room, under bright lights. As we began recording, Long's line of enquiry got me hot under the collar. She asked me about the incident with my mum, which I had detailed in the book, but somehow it felt distasteful and invasive. For some reason, I wasn't prepared for questions of such a personal nature. Had it been a private conversation, I wouldn't have thought twice about going into more detail, but in this moment, aware that there were more eyes on me than ever before, it didn't feel appropriate. In reality, she was just doing her job. It began to dawn on me throughout the course of the discussion, not only the true scale of public interest in my story, but the particular aspects of it that interested the media most. I had only just come from an interview with the *Guardian* and would imminently grace the cover of its popular cultural supplement. This was the sort of generous, comprehensive coverage any artist, writer or campaigner would dream of. Yet there I was, already on the defensive, anxious I had said too much, pondering the consequences or reprisals, while desperately attempting to put the toothpaste of my personal trauma – which I had willingly if naively unleashed – haphazardly back in the tube. I recall thinking of my family, and the fact my raised profile would reverberate in their lives back home. I thought of the inevitable criticism I would receive, the gossip my rising star would generate, and the expectations others who identified with my story may develop with regards to me becoming a public voice on these issues. Of course, when I was writing the book, I had no idea it would go on to do as well as it did. The book is clearly written from a local vantage point and intended to be read by other local people. But before I even returned to Glasgow, production companies were enquiring about the rights to turn my story into a drama for television – offers which I did not explore any further – and within a year, the book would be translated into no less than seven languages, including Japanese and Mandarin. There was a time when I would have concluded that media interest in the more sensational aspects of my book was excessive and exploitative. And that some of the fallout in my personal life, and the lives of people connected to me, was proof positive that sections of the press ought to strive to become more trauma-informed. A time when both these conclusions would have felt not only right to me, but to a large section of my readership. The problem with that though is that in identifying the media as the problem, I rob myself of the opportunity to examine the role I played in the consequences. After all, I wrote the book. I signed up to publicise it. I did the press tour. Nobody was holding a gun to my head. Yes, it may have felt at the time like I had no other choice and that so much was riding on it all going well (my family's security, my own career) but let's be frank here: people provide for themselves and their families every day without going to the very public lengths I eventually went to. The truth is, a part of me needed to be seen. Part of me wanted the validation of winning. I set out on this course with a vague awareness that success, as remote as that possibility seemed, may produce circumstances which may overwhelm me, and despite my reservations, I proceeded, nonetheless. I believe anyone would, in my circumstances, have done the same. The pay off if it went even half as well as I had ever dared to imagine, was simply too great for someone of my modest means to resist. To this day, the consequences of my actions flow from those events. Not all of them are negative, obviously, but had I possessed a greater insight, not simply into what may have transpired had my plan come off beautifully, but also into my own instincts, flaws and vulnerabilities, then I may have done some things differently. In a court of law, ignorance is no defence. I believe the same is also true in the court of public opinion; when we seek to meet our needs, whatever they may be, by appealing to the public square, we must understand that while the spoils may be sublime, they will always be offset by adversities we may have foreseen had we taken a moment to consider them. This is doubly the case when we have no real idea what is driving us to seek this kind of fleeting validation. To be blindsided by our very own nature is to leave ourselves extremely vulnerable, whatever we disclose publicly.

*

As we approach the final third of this journey, it's time to confront the most sacred of all trauma narratives: your story is not necessarily yours alone. Of all the questionable ideas reverberating around the Trauma Industrial Complex, the notion that storytellers reserve all rights to their stories is easily the most ill-considered. To demonstrate what I mean, let us now perform a quick thought experiment for the benefit of those who've swallowed this dangerous idea wholesale. Picture in your mind someone you know for a fact hates you, and for good reasons. Someone you fucked over. Someone you hurt, upset or even abused in the past. Maybe a former employer you let down or stole from. Maybe an ex-partner you lied to, or a kid you weren't there for due to being consumed by your own adversity. Now imagine I go and find this person who really doesn't like you, I mic them up, shove a camera in their face, and ask them to describe in detail precisely why you are the worst person in the world – ever – before uploading the footage to TikTok. This goes for all of you, not just the storytellers. Really cast your mind back to a time when you could safely say you were the villain of someone else's story rather than the hero of your own. And now imagine that darker, less caring, more selfish iteration of you was broadcast publicly for everyone else to judge. That chapter in your story you haven't integrated into your online persona. That part of your past your gushing colleagues and loyal friends know little about. Or maybe you're still that person and haven't been found out yet. It doesn't bear thinking about, does it? How would you react in such a scenario? Would you grin, bear and then grudgingly affirm your sworn enemy's version of events? Would you commend their bravery before deleting all your social media accounts? Or would you do what most people do when confronted by the overpowering scent of their own bullshit, and contest, deny or attempt to contextualise the slanderous accusations.

Now please be aware that I am not saying perpetrators descrve a right of reply before stories of their abuse are told – that's ridiculous. What I am saying, however, is if safety is valued as much as the specific details of the story, then we might wish to consider the potential blowback resulting from testimony which receives such a level of prominence, that others connected to the story get wind of it. We are all entitled to our own interpretation of our story, but should we decide to share our version of events with a wider audience, however small, we must be prepared for consequences. Not least, if aspects of our story are disputed, or told from a place of woundedness. While stories of trauma centre victims and survivors, for good reasons, the narratives invariably involve other people who may or may not be completely comfortable with such private matters being shared so publicly – and not all of them are the villains. People we love, even people who helped us, may also experience feelings of vulnerability or even trauma, as a result of a poorly considered disclosure which breaches their personal boundaries. While our intentions might be noble, and it may indeed transpire at some later point that putting it out there was of some value either to ourselves or others, trauma is an unpredictable force in the world, one which must be handled with utmost care – easier said than done in the twenty-first century. The means to share aspects of our private lives publicly have never been more accessible. Devices we carry in our pockets now come equipped with tools that were once the exclusive domain of newspapers, broadcasters and filmmakers. Apps like WhatsApp, Snapchat and

Telegram allow us to communicate instantly, while platforms such as YouTube, Facebook, X and TikTok go further, enabling us to broadcast our lives to potentially millions of people. This surge in connectivity is rooted in our innate desire to connect, to express and to share our experiences. It has birthed an industry that overshadows almost every other, reshaping how we think, speak, feel and act. Increasingly, we view our lives as content and ourselves as brands, worth documenting and disseminating for others to see. When I was young, we marvelled at the lives of others – actors, musicians, public figures. Today, that gaze turns inwards. We are no longer just observers but active participants, crafting and starring in our narratives. The tools at our disposal have made us all publishers, and many of us now feel compelled to share our stories, believing we have a right – even a duty – to do so. The stories we choose to tell today often push beyond the limits of what was once deemed appropriate. It's no longer just about sharing the good news of a new job or a family holiday; increasingly, people disclose their opinions on divisive topics, their struggles with health, and even the intimate details of their relationships. The taboo lines of what should remain private and what is fair game for public discussion have been redrawn. A generation has grown up in a world where participation in this tell-it-all culture feels not just encouraged but necessary. The compulsion to share, to be seen and heard, becomes a defining feature of modern life. But what happens when the story involves more than just ourselves? What happens when the act of putting it out there impacts others – those who may not have consented to having their lives broadcast in the public domain? Of all the stories shared publicly, few carry as much weight – or risk – as those rooted in trauma. Narratives of pain and adversity resonate deeply with audiences, creating connections that feel meaningful. But trauma is rarely a solitary experience. It often intertwines with the lives of others – family, friends, acquaintances, even antagonists. When we tell these stories, we often do so from a place of conviction: it's our trauma, our story, and therefore ours to share. But this perspective overlooks a critical truth: our experiences, no matter how personal, rarely occur in isolation. Disclosures, particularly those involving trauma, ripple outwards in ways we cannot always predict. They may spark feelings of betrayal, vulnerability or anger in those who see themselves reflected – or excluded – in our narrative. Publicly sharing trauma can leave us open to unexpected consequences, both external and internal. Often, facilitators and consumers of our stories are unfamiliar with these because these consequences are not always a feature of the front of house production – they are managed behind the scenes if at all. Too often, blowback seems to be reacted to rather than foreseen and mitigated. Disclosures occur, and then the damage limitation begins – see *Baby Reindeer* again, which appears to be a prime example, or the story of Steven Dymond, who took his own life after appearing on *The Jeremy Kyle Show*. While Dymond had a history of suicide attempts and an inquest into his death concluded that there was 'no causal link' between his appearance on the show and his death, Dymond was reportedly 'distraught' and 'broken' after returning from filming. The inquest heard he was 'booed' by the audience during the show after a lie-detector test suggested he was not truthful when he denied cheating on his partner. Kyle said he felt 'exonerated' by the inquest finding and ITV stated that the inquest showed the show had 'comprehensive duty of care processes covering the selection of contributors who appeared on the show and their care both during and after filming'. Nonetheless, the clamour to platform a great story may take precedent over the safety of the person telling it, or indeed, the people connected to it. We must also consider more carefully the various ways a story may come back to bite us, particularly if we are living with active trauma. The mere act of speaking up in the public square invites scrutiny. It then stands to reason that living part of your life in that kangaroo court will put even the most resilient nervous system through its paces. How will you cope when you realise that for every compliment you receive, there's an insult? I've overheard people talking about me outside venues, or in the corridor by my dressing rooms while on tour. Most people are kind and considerate in person, but when they think you're gone, their true feelings are expressed without fear or favour. It can be very painful and confusing when faced with ungenerous critiques of our stories and even ourselves, or evidence of disloyalty or two-facedness. Our words, our intentions, even the validity of our experiences can be questioned. Some people will challenge the accuracy of our account, while others may take issue with our decision to speak out at all. This can be particularly distressing when the criticism comes from those close to us. Family members or friends may feel misrepresented or hurt by how they are portrayed – or if they have been omitted entirely. Even those who were supportive during difficult times might resent the lack of recognition or feel aggrieved by the way certain events are framed. In addition, members of the public, emboldened by the disinhibiting effects of social media, can be cruel and unforgiving. Behind the safety of a screen, people may hurl accusations, insults and threats that cut deeply. I've received death threats, threats of violence, threats of being reported to authorities for crimes

I did not and would never commit. I've experienced long periods of depression and anxiety, afraid of being attacked both physically and verbally by strangers who appeared particularly aggravated by my prominence. For those of us whose sense of self has already been eroded by trauma, these comments can feel devastating. Without a strong internal foundation, and a robust support network, we risk internalising these judgements and starting to believe the horrible things said about us and our stories. The fallout can extend beyond hurt feelings. Some may actively retaliate against us, contesting our version of events or presenting their own counter-narrative. Former abusers, in particular, may seize on the opportunity to reassert control or inflict further harm. They might use legal threats, public denials or personal attacks to silence or discredit us, perpetuating the cycle of abuse in a new, public arena. Disclosing trauma can also trigger internal conflict. The act of sharing may reopen wounds, activating unresolved pain and leaving us vulnerable to feelings of regret, shame or exposure. Once the words are out there, they cannot be taken back, and the permanence of a public testimony may eventually weigh heavily. It's tempting to view our stories as solely ours, but the reality is more complex. Every story has multiple perspectives, and the truth is rarely absolute. Memory is fallible, shaped by emotion and time, and no version of events can ever be entirely objective. This is particularly true when considering the theory of false memory. Psychological research shows that our recollections are malleable, influenced by external suggestions, emotions and even the act of recalling events. What feels unquestionably true to us may not align with how others remember it – or how it objectively happened. This is particularly true when our story involves others – especially those who may not share our interpretation of events. Again, let me reiterate that this isn't about centring the feelings of perpetrators or appeasing those who are misinformed or who may have wronged us. It's about recognising that our safety, and the integrity of our narrative, depend on carefully considering how public storytelling might reverberate in our personal lives, our immediate communities and wider afield.

Every consequence I've described is one I've lived. I've faced legal action, accusations of fabrication, and strained relationships with people I care about. I've been called a liar and even accused of selling my own mother out by making up stories about her for money and fame. Words I can never take back, disclosures I can never undo – these weigh on me, even years later. I've learned the hard way that trauma, when shared thoughtlessly, has a way of snapping back. It can undermine the sense of catharsis or connection we seek, leaving us feeling exposed and vulnerable. And despite the harms we endured, we must accept that we, too, are capable of causing harm, particularly when we prioritise our need to tell our story over the impact it may have on others. In many ways, this book stands as an attempt on my part to redraft a fuller story, which reflects on the broader consequences of disclosure and acknowledges the challenges its prominence thrust not only upon me, but on my family and friends. To tell your story responsibly is to acknowledge its complexity. It's to recognise that while you have a right to your truth, others have their truths too. It's to accept that sharing trauma publicly carries risks – not just for you, but for everyone connected to the narrative. Trauma, as I said, is a volatile force. When handled with care, it can foster understanding and healing. But when wielded recklessly, it can snap back – at yourself, at others, and at the very fabric of your relationships. In choosing to share, or to facilitate another's testimony, we must weigh these consequences carefully. Because while our stories are ours to tell, they are rarely ours alone.

'People are trapped in history
and history is trapped in them.'

James Baldwin,
The Fire Next Time (1963)

CHAPTER 12

The Story of Us

*Is all of our identity fixed, or
can some of it be rewritten?*

Identity, like stories of trauma, is deeply personal and endlessly complex. It's not just one thing; it shifts depending on how you look at it. And like trauma, it's tied to the most vulnerable parts of us, shaped by our experiences, relationships and the world around us. Psychologists might define identity as our self-concept – what we believe about ourselves, our values, and how we see our story unfolding over time. Sociologists, on the other hand, might argue that identity isn't just internal; it's a social construct. It's about the roles we play, the groups we belong to, and how those broader systems shape us. Cultural perspectives take this further, framing identity as something shared: traditions, language and heritage that connect us to a collective past. Then there are postmodern thinkers like Foucault, who throw a wrench into the whole idea of fixed identities. For them, identity is fluid and ever-changing, shaped by power structures and the narratives we're immersed in. Intersectional approaches layer this even more, showing how identities – like race, gender and class – overlap in ways that shape unique experiences, often tied to inequality and struggle. Identity doesn't exist in a vacuum – it's constantly reflected and refracted through how others see us. This external validation, or lack thereof, can leave us feeling recognised or erased, connected or isolated. And how could we forget the cold, hard politics of it all – identity is also about power. Who gets to define what an identity means, who belongs to it and who benefits? These questions often highlight long-standing inequalities and create ongoing struggles for recognition, fairness and visibility. What makes identity so personal, and often so contentious to discuss or debate, is that it sits at the core of how we see ourselves and where we feel we belong in the world. It's inherently emotional. Which is why I pose the following question with a certain trepidation: when it comes to who we believe we are, on the inside, and not necessarily our immutable traits or how we are legally defined, how much of our identity is fixed, how much of it is a story, and how much can we rewrite? Questioning someone's identity can feel like an attack, not just on their beliefs but on their very existence. For people whose identities are tied to histories of oppression or trauma, these conversations can bring up deeply painful memories. So, let's begin with my own identity as a case study. I don't know if you've picked up on this, but I'm working class – hardly ever mention it, right? Class is probably the defining facet of my identity, the thing people most associate with me when I speak publicly. Whether praised or criticised, the comments I receive are almost always tied to something I've said about class inequality. It's as central to my identity as trauma is to my nervous system. And there's a good reason for that: I have consistently emphasised this aspect of my identity above every other. The fact I'm male, white, heterosexual, Catholic, an addict, an introvert, an artist, a house-husband, a father, a son, a brother, a gym-bro, a lover of slippers, a prolific daytime napper, a purveyor of the finest scented candles – these facets are never telegraphed with the same intensity as my class background, if at all. Indeed, these other facets are often seen through that lens – adjuncts to my working classness. But there are many people from my class background who see no value in viewing themselves or the world through the lens of class. For them, despite growing up in similar economic circumstances, and sharing similar material interests, a class-based identity is not one that interests them. Sometimes, this goes beyond mere ambivalence; the notion of class is rejected forcefully, seen as a self-limiting belief, or a value judgement on others. While I struggle to understand how anyone could minimise or dismiss a reality that I consider material in basis, their lives don't appear adversely impacted by the absence of a class-based identity. That is to say, they don't experience consequences any greater than those who do hold the class analysis to be true. For me, on the other hand, well, this root belief has defined not only how others see me, but how I see myself. Truth be told, this facet of my self-image has come under immense strain in recent years, not least when I began earning more money following the success of my debut book. Overnight, it felt like so many aspects of my life changed. The people I interacted with became gradually less representative of where I grew up, and more representative of the spheres I'd spent my youth calling out and criticising. The nature of life's dilemmas also began to evolve; meeting my basic needs was no longer the daily conundrum – sifting through the vast opportunities I was suddenly being offered became the central preoccupation. I recall taking a friend out to dinner and attempting to open up to them about some of the unforeseen challenges that come with a sudden increase in wealth, status and visibility. The look in their eyes of pure disgust was hard to ignore. They didn't have to say a word; I felt their resentment radiate across the table. This wasn't resentment about what I had that they didn't, it was deeper than that. My problems were not in fact real problems. My problems were the sort of dilemmas the average person would count themselves lucky to have: the problem of being overwhelmed with well-paid work; the problem of ascending suddenly to a higher tax bracket while financially illiterate; the problem of getting recognised; the problem of being publicly noted and your work being seen, discussed and critiqued. I was now sleeping with the enemy. I was becoming middle class – a fate worse than death for a self-conscious class warrior like me. When your whole identity is rooted in the notion of being a have-not, who must fight for opportunities hoarded selfishly by poshos and elites, life becomes very confusing very quickly, the day you wake to discover you have everything you need and more. Throughout this period, rather than reassess the legitimacy of my claims to being working class, and make the necessary adjustments, I instead doubled down, self-consciously, hesitant to admit to myself or concede to anyone else, that my class experience was evolving. I wrote in a self-flagellating way about feeling embarrassed travelling first class. I used my documentary films to re-telegraph my class position; threatening to punch a butler, pointing a loaded rifle at a landowner – stunts that reinforced the notion I was still a radical who'd somehow wangled their way onto national television through sheer Machiavellian skill and strategy. But the mere fact I was even given a television series in the first place was surely an indication that despite my pretensions, my radical flame had dimmed ever so slightly. Not necessarily because my politics had changed, but because my priorities had. I was now a father, an addict in recovery and approaching middle age. I was doing what most young radicals do in the end – I was growing up and making space for younger campaigners to emerge and ruin everything. But part of me just couldn't let go. The sneering remarks I'd read or hear from the true faithful (usually activists years my junior with no kids, few responsibilities and little clue about life outside university libraries, direct actions and the lower tiers of trade unions) bothered me intensely. The fact I now had a growing audience that valued and supported me, that didn't determine my value based on whether they always agreed with me or not, but on how I considered issues and expressed myself, was nowhere near as important to me as proving the resentful doubters wrong. My second book was in many ways my attempt to very publicly re-nail my political colours to the mast; an onerous tome exploring every facet of class inequality as I saw it, in exhaustive depth – to prove I still cared and hadn't turned my back on my people. A book more than twice the length of this one, in which the only personal testimony the reader gets is me either reminding them how poor I used to be or describing how uncomfortable I was, moving in the middle-class world of media executives, publishers and book festivals. Funnily enough, few of the radicals I'd hoped to appease with that book actually read it, many who did sneered at it, and much of my wider audience who gave it their time, found it informative and well written but difficult to connect with – and no wonder. The lengths I've gone to, in my desperate hope of clinging to my working classness, as if accepting my life has changed would represent some kind of betrayal, reveal the pitfalls when identity becomes a performance rather than an extension of who you really are. What strikes me now, as I reflect on this more openly and with less ideological rigidity than I would in a book about inequality, is that both class and trauma were identities I didn't exactly choose – they were handed to me. This is what is called in sociology an 'ascribed' identity. It's a facet of our identity we have very little say over. This contrasts with an 'achieved' identity, when we internalise new facets previously foreign to us, in which we play a more active role – a profession, becoming a parent or taking up a hobby like sport. Trauma was inflicted, the working class is a group I was born into. Sure, I've leaned into both at times, consciously or not, and hold both dear to my heart for different reasons, but these weren't concepts I sat down and decided to adopt. They were roles I was assigned. The most appealing of the range available to me. Yet I've clung to them both so tightly over the years, like prized possessions, and the complications arising from them, despite having no real ownership over their place in my life. No matter how you define identity – whether it's tied to your self-image, your culture or an intersection of influences – you've got one and it isn't always rooted in the truth. The narratives we tell ourselves about who we are aren't purely factual; they're stories designed to keep us afloat, that make the choppy seas of life easier to navigate. We lean on selective memory, reinterpret events and idealise situations to protect our sense of self. We wear masks to fit into social moulds, shaping identities that feel acceptable but aren't necessarily always authentic.

On top of that, the stories we inherit – from family, culture or media – play a huge role in how we see ourselves. These collective narratives often overshadow our individual truths, and when they clash with our experiences, or how we'd like to be seen, we may suffer discomfort. We might tweak our identities or craft justifications to ease the tension. While this can offer short-term comfort, it may also obscure deeper truths that we'd benefit from confronting. Of course, there are parts of our identity that aren't up for debate. Your ethnicity is non-negotiable. Your sexuality is a fact of your life. You can't think or meditate, or manifest out of those realities, just like you can't actualise your way out of oppression, persecution or war. Sometimes, external forces in politics or media place immense pressure on these immutable aspects of who we are, and the fight to simply exist as ourselves becomes the basis for necessary political struggle. These musings are not an attempt to downplay or dismiss that reality. Beyond those material facts, however, much of identity formation is highly selective. We highlight certain traits while leaving others in the shadows. When hip-hop became central to my identity – something we'll explore in the next chapter – I leaned into a persona of toughness and readiness for conflict, even though I'm naturally mild-mannered and empathetic. In my element, I was the person at an afterparty making sure everyone had enough water, was warm enough, and knew they were welcome to lie down in my bed if they needed a rest. But in my rap guise, these traits were dialled down. I was often dismissive, sarcastic and aggressive – a bully, basically. That version of me felt more like a suit of armour than a reflection of who I really am and when I acted out on those impulses, consequences always followed, internal or otherwise. How many of you have done the same – crafting a version of yourself that feels like survival gear but isn't really 'you'? How many of us waste time and energy trying to maintain identities that no longer fit, or no longer serve us? At its core, identity is about figuring out who we need to be to survive, belong or matter. But too often, we wear these identities like ill-fitting outfits – surface-level statements that only tell part of the story. We rarely stop to think about where these identities come from or what they're made of. Worse, we construct additional stories to justify the personas we cling to, layering distortions and fabrications on top of each other, hoping the internal dissonance will pass. Like many of you, I was unable to quiet that nagging sense of living out of the reality and took to self-medicating with alcohol, drugs, overworking and consumerism. Each unhealthy coping strategy, deployed to reinforce the walls of the illusion, shutting reality out for just long enough to get me out of a tight spot. The truth is much of who we become unfolds by chance; random events we retrofit with stories that create the impression we had some guiding hand in it all. Many of the things we hold central to our identity are in fact based on mimicry, guesswork or circumstances of birth. And yet we act as though traits – things we didn't choose – are personal achievements, while ignoring the aspects of our identity that come from our own agency, like how we treat people or our private struggles. It's no wonder so many of us, weighed down by trauma, feel unsteady when life inevitably tests us. What has also begun to fascinate me, though, is that there are parts of my own history my memory cannot touch. The stories of previous generations and the hardships they endured in a world vastly different from the one I grew up in. Aspects of my history which likely play as significant a role in how my life has unfolded as any other factor – including my relationship with my mother. Strangely, however, these important details have never found a place in my story, nor my identity, which may suggest that despite how tightly we often cling to and assert it, aspects of our self-image can ring as hollow as some of the stories we tell ourselves about our lives. Well, here's a story you've never heard from me before. A story I never thought to tell because it hadn't been told to me. The story of my paternal grandparents, and how a culture where trauma was rarely acknowledged or discussed left them trapped in assigned identities until the day they died.

*

My granny's name was May. Every Thursday during the summer holidays, we would walk from her house on Leithland Road, Pollok, to the post office at the Pollok Centre to collect her pension. It was a weekly ritual for both of us. For her, it meant she didn't have to steal money from the inner lining of my grandfather's coats. For me, it meant sweets, computer-game magazines, and sausage, beans and chips at the café overlooking what was then a modest-sized Tesco – if you can imagine such a thing. Before we did the shopping, though, we often had two other stops to make. First, we'd buy our *Big Issue*. At the time, I was too young to understand the significance of this ritual. Still, we always stopped to chat with the vendor before purchasing the week's edition. Sometimes, my granny would even take the magazines from him and sell them herself, giving him a chance to get a cup of tea and something to eat. Our next stop was outside the Pollok Centre entrance, where activists regularly gathered, handing out flyers on everything from the Militant movement – still strong in the political vacuum left by Thatcherism – to school-closure protests and campaigns against the M77 motorway led by the Birdman of Pollok, Colin MacLeod. My granny knew many of the activists by name, and they knew her too. She commanded respect, a quiet moral authority you wouldn't dare cross. The sort so common in working-class communities, where people make their values visible through their actions rather than proclaiming them remotely from digital rooftops. She was the head of our family. One word from her, and everybody fell in line. Some of my earliest memories are of protests by her side. In these tight-knit activist communities, I became known – not because I did anything significant in the beginning, but because I was always with her. It mattered to my granny that I understood what was going on in our community: its history, its struggles, and the sharp decline that had come to define it. She wanted me to see the forces shaping the world around us – not just for their impact, but for what they revealed about resilience, solidarity, and justice. My granny's interest in local politics extended beyond protests. She was a prolific letterwriter, a woman who understood the power of words. If someone crossed her or a wrong needed righting, out came the notepad and fountain pen, steady in her left hand like a maestro wielding a baton. As she leaned into the page, thoughts flowed effortlessly, precise yet expressive. In those moments, she became someone else entirely – self-assured, focused and utterly immersed. I realise now that my granny was probably a writer. A writer driven by duty and fuelled by the desire to right wrongs, acknowledge kindnesses and demand accountability. But she was also a working-class woman, born into a time and place where writing – like so much else – wasn't seen as a viable path for someone like her. What could she possibly have to say? Plenty, I suspect, had she been encouraged to pursue it. My granny, like so many women of her generation, never had the chance to fully embrace or explore her true identity, hemmed in as she was by the cultural expectations of her time, shaped by the grinding poverty of the Great Depression and then the Second World War. Growing up in Glasgow during the 1930s and 1940s wasn't easy, not for anyone, but especially not for children. Life for working-class people was tough at the best of times. Families crammed into crumbling tenements, with damp walls and cold floors that no amount of blankets or fires could ever make comfortable. Food was always in short supply. It wasn't uncommon for illness to sweep through entire streets, and when your parents were already struggling just to put bread on the table, there wasn't much to do but get on with it. Even in the harshest of times, however, people found ways to carry each other through. When the war came in 1939, it was as though the city's already fragile existence was put under a magnifying glass and left to smoulder. Glasgow was an obvious target for German bombers, with its shipyards and factories making it a critical part of the war effort. The Clydebank Blitz of 1941 was a trauma that left scars on everyone who lived through it. Hundreds of people lost their lives, whole streets were wiped out, and thousands were left with nowhere to go. For children, the terror of those nights was seared into their nervous systems. The sound of air-raid sirens would cut through the evening, sending families scrambling into Anderson shelters or huddling in subway stations. The earth shook and the sky lit up with explosions. When it was finally safe to emerge, the scenes were devastating – homes reduced to rubble, neighbours grieving loved ones, and familiar streets unrecognisable. That kind of fear doesn't quickly subside. It lives deep in your body. And yet, through all that fear and loss, something remarkable happened: people came together. It was the women of Glasgow who led the charge, holding families and communities together with a strength that's hard to overstate. Mothers, aunties, grandmothers – they became the glue that kept everything from falling apart. They found ways to make something out of nothing, sharing what little food and resources they had, pooling together to look after children, mend clothes or offer a shoulder to cry on. These women didn't just keep things running; they created a sense of hope, however faint, in a bleak war-torn world. They were the ones who taught kids what it meant to care for each other, to lean on your neighbours, to survive not alone but as a community. Rationing made life even harder, and hunger was a constant companion. But Glasgow women were resourceful, turning scraps into meals and making sure that no one went without. A pot of soup stretched across families, a neighbour shared their last bit of sugar, a child offered a piece of rationed chocolate – it was these small, everyday acts of kindness that reminded people they weren't alone, soothing the effects of trauma reverberating around their lives. Even in the worst of times, the downtrodden found ways to lift one another up. But the trauma lingered, as there seemed little reprieve from this tough existence. Children inevitably soaked up the fear and grief around them, even if they couldn't fully understand it. The sight of their parents breaking down, the loss of neighbours and friends, the knowledge that nowhere was truly safe – it left marks that couldn't be seen but were felt for a lifetime. For those children, the war wasn't just something they lived through – it shaped who they were. My granny was born into a world where being a woman, particularly a working-class woman, meant a life defined by duty, sacrifice and survival. Her identity was shaped not by choice but by circumstance – by the weight of social and cultural norms that assigned her the role of a Catholic housewife, a mother and a caregiver. And while she fulfilled those roles with immense strength and grace, and there is great virtue in anyone who performs them, it's impossible not to wonder what might have been if she'd lived in a world that allowed her to pursue the fullness of who she was. She wasn't given room to cultivate those parts of herself. Instead, she was directed by the currents of expectation along the path of family, faith and community. She was assigned an identity which she resisted at points but learned to grudgingly accept. And while those of us who knew her saw flashes of the greatness that she surely carried within her, it is sobering to think of all the other grannies, mothers and women like her – women with the capacity for brilliance – who were quietly stifled by the world they were born into. Women who were conditioned to embrace, celebrate and assert identities they had very little hand in cultivating. My granny was so much more than her ascribed identity. She was a writer without title, an activist without platform, a woman of intellect and emotion whose true potential was sadly never realised. Her identity, rather than a badge of honour, was often an escape room she gave up trying to get out of. Even as tough as she seemed to me as a child, her true origin story sometimes slipped through. She told me stories of being punished at school for being left-handed, her teacher forcing her to show her hands again after she instinctively pulled them away in fear of the belt. She spoke of her father, a Second World War veteran slowly driven mad by a bullet fragment lodged in his brain, who would sometimes look at her with frightening rage as though she were a stranger. There were darker stories too, alluded to but never fully told, of male abuse and mistreatment and the ways it marked her life.

Nelson Mandela was a big hero of hers and his memoir, *Long Walk to Freedom*, sat by her bedside for years. She kept a scrapbook filled with clippings, letters and mementos that revealed the breadth of her interests and a sentimental streak which perhaps was harder for her to admit to. The centrepiece of this scrapbook was a section on the Hillsborough disaster – pages packed with photographs, testimonies and reports. For visitors, it was an emotional read and many would weep as they worked through it during their visits. Her connection to Hillsborough was twofold: an assault on the working class could not be ignored, and she lived for football. A die-hard Celtic fan, she took me to matches every other week until her health and finances no longer allowed it. Back then, I didn't understand the values all these things reflected about her. I just wanted her to hurry up. I thought she gave people too much of her time and attention. My granny shaped my politics more profoundly than anything else. It wasn't the dire material conditions of Pollok but how she carried herself that made me a socialist. I equated her politics with her values and how she lived her life. It wasn't an innate sense of morality that made me care about vulnerable people; it was her example. In the 20 years since her passing, I've learned it's who she was that mattered. Not everyone who claims her politics conducts themselves in the manner she did. But beneath her local persona as a fighter, ever ready to stand her ground, she was also repressed and even submissive. She admired changemakers and anyone who fought against oppression and injustice yet struggled to assert her own needs and basic rights, even in her own marriage.

My grandad's name was Tommy. He was a respected but difficult man. An Irish immigrant who came to Scotland looking for work in the 1950s, he encountered my granny in a public park on Glasgow's South Side; taken by her beauty as they passed one another, he performed a vital check to see if they were compatible – throwing his rosary beads on the grass and gauging her reaction. When she didn't burst into flames at the sight of them, he concluded they were soulmates. They courted, fell pregnant and had the first of five children in 1958. My grandad, like her, carried deep wounds from childhood. For children born in 1930s Ireland, in the newly formed Free State, the legacy of British imperialism was more than history – it was a presence that hung in the very air they breathed. Their parents and grandparents carried the scars of colonisation, the War of Independence and the Civil War, passing those wounds down in ways both spoken and unspoken. These children inherited a world where the fight for freedom was celebrated, but the cost of that freedom – partition, poverty and division – was a daily reality. Life for these children was marked by chronic hardship. The economic strain of the 1930s, worsened by the 'Economic War' with Britain, meant that many grew up in rural poverty, with little opportunity for upward mobility. Parents struggled to make ends meet, often reliant on subsistence farming or seasonal labour. Hunger was a familiar feeling, and emigration loomed as an eventual inevitability, with older siblings or neighbours leaving for Britian or America to send money home. For these children, the dream of a better life often meant separation from the only home they had ever known. The cultural revival underway in Ireland added another layer of complexity to their lives. At school, they were taught to speak Irish, a language that had been suppressed under British rule but was now being reclaimed as central to their identity. Yet for some, this effort felt forced and disconnected from their everyday realities, where English dominated. Education itself was often strict and authoritarian, heavily influenced by the Catholic Church, which had stepped into the role of moral and social authority after British rule. Corporal punishment was common, and children were raised with a sense of fear – of God, of authority and of stepping out of line. Many cases are documented of excessive use of the cane or rod, for example, and of parents lodging complaints with authorities concerned about alleged abuse that were often dismissed or minimised. The church's grip extended far beyond the classroom, shaping not only their moral upbringing but also their sense of worth, shame and place in the world. The shadow of the British Empire lingered in subtle but powerful ways. While the Free State had broken away, many of the structures left behind – legal, economic and cultural – remained. Children might hear their parents speak bitterly about landlords or land agents, reminders of a time when their families had been tenants on their own soil. Others would grow up hearing stories of relatives who had fought and died in the fight for independence, stories of unspeakable violence told with pride but also with grief. Partition was a wound that bled into their understanding of identity. These children were Irish, but they were also growing up in a fractured country, with an invisible border dividing not just land but the sense of what it meant to belong. At home, conversations about Britain often carried a mixture of resentment and pragmatism. While British rule was denounced, the reality was that many families still depended on Britain – whether through trade, money sent from loved ones who'd gone there for work, or even the jobs that awaited those who would inevitably leave themselves. For many children, this created a confusing and contradictory dynamic all too common in post-colonial communities and one that mirrored abuse dynamics at the root of so much trauma: Britain was both the oppressor and the opportunity, the enemy and the escape. Britain was the perpetrator and the rescuer. This ambiguous ruling force was mirrored by the Catholic Church, which gave communities a sense of moral structure and guidance, but it was achieved through overbearing and unaccountable theocracy. For some, the inherited trauma of colonisation was compounded by the silence that often surrounded it. Parents and grandparents who had lived through famine, war and loss rarely spoke of their pain openly. The weight of those unspoken experiences settled over the home like a fog, shaping how children understood struggle, resilience and the role of suffering in their lives. Emotional expression was not widely encouraged, and many children learned early on to carry their burdens quietly, developing survivor-personas which often accompanied hard-drinking. The stoicism that had helped previous generations survive now became a barrier to processing trauma, leaving many children to grow up with a sense of unease they couldn't name but that never quite left them.

My grandad was one of those children. Despite having every reason to share in my granny's socialist beliefs and distrust of the British state, he adopted a surprising worldview – despising socialism and socialists, while embracing the opportunities of a life in the UK without much protest. This, of course, was not a wholehearted embrace, but one born of necessity. For him, life was simple: you went to work, made your money and either came home at the end of the day or went to the pub. My enduring memories of my grandad are mixed. I came to realise later in my life that he was deeply troubled and had great difficulty expressing emotion – something alcohol helped and hindered in equal measure. He never missed a day's work as a plasterer, nor a Sunday mass, and you could always count him in for a night's drinking at any one of the locals around his end of Pollok. There, he found a brotherhood among characters of varying quality, having mastered the art of blending in – a skill central to assimilation. On Friday nights, not long after last orders, you'd hear his key in the door and his slow, heavy footsteps along the hall towards the living room. A clink of the glass bottles of fizzy drinks lining the hall indicated when old Tommy was, perhaps, less steady on his feet. His arrival would cause those of us at home to brace ever so slightly, wondering which version of him were we getting that evening. On a good night, he'd poke his head round with a mischievous drunken grin on his face, before producing a polythene bag full of sweets which were emptied onto the carpet for the grandkids to devour. At just the right level of drunkenness, he was hilarious company, though admittedly this wasn't always intentional. He wasn't an educated man, and so he'd frequently mispronounce common words. This, coupled with his thick Irish accent, was beyond comical and I even suspect that sometimes he played it deliberately for laughs – he wasn't as dim as he often wanted or needed people to believe. His drunken ramblings were like routines, featuring the same flights of nostalgia, often punctuated by song. 'The Fields of Athenry' by Pete St. John was a family favourite. Set against the backdrop of the Great Famine in the 1840s, the lyrics follow a fictional man from near Athenry, County Galway, who steals food to keep his starving family alive – only to be sentenced to transportation to the penal colony at Botany Bay. The song has become deeply woven into Irish cultural memory, resonating with generations as a haunting reminder of famine, displacement and the legacy of British colonial rule. Only through the lyrics, and those of similar ballads, did I ever get a glimpse into what truly made him tick – or rather, what he may have been suppressing.

> *By a lonely prison wall*
> *I heard a young man calling*
> *'Nothing matters, Mary, when you're free*
> *Against the famine and the crown*
> *I rebelled, they cut me down*
> *Now you must raise our child with dignity'*
> *Low lie the fields of Athenry*
> *Where once we watched the small free birds fly*
> *Our love was on the wing we had dreams and songs to sing*
> *It's so lonely round the fields of Athenry.*

Those were the better nights. The ones I think of most fondly. On bad nights, we were grateful if he fell asleep furious in his chair after one of his verbal tirades. He wasn't an affectionate man by any stretch but he showed love through acts of service; handing his wages over at the end of the week without fail, tending to repairs in the house, loaning people money when they needed it. That said, he could surprise you. Sometimes, you'd get a sniff of a gentler man buried beneath all the avoidance. Someone who wanted to connect with people around him – even Rangers fans. His occasional observation that some guy down the pub was 'all right for a Protestant' was often the source of laughter – because he was deadly serious. My grandad's battle for authenticity was no different from my granny's, though his was fought on a different internal front. He had few emotional outlets and lacked the human touch and natural charisma she possessed in spades. His interests seemed narrower, to me at least, though he mellowed out a little as he got older. After he retired, he spent more time at home – much to my granny's irritation. He loved Westerns and watched them religiously, even learning to operate a VHS player and replaying them on tape. Without fail, at the end of every film, as the story reached its emotional climax, he'd clumsily reach for his newspaper and hastily open it, though not to browse its pages – he was using it to disguise that he was weeping. We'd all have to pretend we couldn't hear him sniffling. It wasn't part of our family culture to ask my grandad if he was OK. Men of his vintage didn't need emotional support, apparently. In truth, he was a deeply sensitive man who could not express feelings without the aid of alcohol. This inability to express love outwardly meant his children were often denied basic kindnesses and affection. Far from being a bad father, he was simply limited in ways that may have lifelong consequences – warmth and affection were not modelled to him with any consistency – though not so limited that he couldn't see the impact. I believe he knew he should have done better. He knew the kind of man he wanted to be – the hero of his Westerns – but he was locked in behind a wall of ascribed masculine roles and traits, under the lock and key of unintegrated trauma. On his shins, thick red sores would flare up at night, which he'd often scratch and tear at with coarse fingernails in drunken sleeps. The sound was excruciating to sit through. As his wounds bled, he'd mutter or sometimes even cry out, 'God forgive me', or repeat, 'I'm sorry', over and over, so utterly tormented he was by fear of a wrathful creator, gleeful at the prospect of his eternal damnation. After my granny passed away, he began to open up more, partly because the whisky had by then become his daily affiliate in her absence. He'd share his thoughts on wildlife documentaries he'd seen, or memories of us as children. He showed interest in my early music, asking me to play him cassette tapes of my very first recordings. 'How do you remember all the words?' he'd ask me – a question I get to this day. And he'd cry openly about his grief in ways that fully conveyed his love and respect for my granny – love and respect she rarely got to hear or experience because he was so closed off when she was alive. He died suddenly within weeks of my granny – a gastric haemorrhage, caused by alcohol – his body discovered by my sister, who kindly cared for him in the final weeks of his life.

*

Both of my grandparents spent much of their lives trapped in identities that didn't necessarily reflect who they truly were. The traumas they carried from childhood were locked in there too, buried beneath layers of expectation and duty, and the fear of stepping out of line. Had you asked them who they were while they were alive, they likely would have given you a long list of attributes, characteristics and roles – some they embraced, others they bore out of obligation. You might think we live in better times now, free to express our needs and desires, most of us – though certainly not all – largely liberated to be who we really are. But is that the whole truth, or just a comforting narrative we tell ourselves? With all the technological and cultural progress since my grandparents' days, have we truly evolved in how we understand ourselves? We certainly have more ways to declare who we think we are, who we want to be, how we wish to be seen. But has this freedom led to a more stable, grounded sense of self, or more involved and harmonious communities? Or have we simply been given more choices in the masks we wear as individuals and groups – disguises designed to mislead others, and maybe even ourselves, about what we actually feel and believe, and what we truly want from our lives? This is why I've grown weary of the cultural obsession with reclaiming, asserting and defending identities. Yes, as I've discussed, there are aspects of identity that inevitably come under attack and must be defended. I'm well aware that the freedoms I enjoy are not universal – not even in the UK – and that despite my many hardships, I have enjoyed privileges inherent to some of my immutable characteristics. But beyond these non-negotiable facets, how many of us actually understand why we are the way we are, or even attempt to grapple with the deeper questions of identity? Today, the pendulum has swung from rigidly conforming to expectations, to curating our identities self-consciously, as if performing them for an audience. But I believe for many of us, that same fundamental sense of duty and repression persists. Trauma discourse often implores us to locate the source of our wounds in the generation that immediately preceded us, but we rarely consider what our parents went through – they were just supposed to know how to meet our needs. As for our grandparents, few of us really grapple with the pivotal role their upbringing, and that of their caregivers, played in how our lives unfolded. When placed in historical context, my own notions of identity start to feel incomplete. Trauma, I've come to realise, casts a much longer shadow over my family than I ever acknowledged, reverberating across generations. Yet for the longest time I attributed my trauma almost exclusively to the nine or so years I spent with my mother before she left, and in the poverty of the 1980s and 1990s. I once thought class politics was my only inheritance from my granny, but it turns out she was a prolific writer, an activist, fond of prescription medication and prone to the odd suicide attempt. My grandad, a figure I never considered particularly influential, I assumed was a proud Irish immigrant fleeing a homeland torn apart by British imperialism. But perhaps part of him was happier to start over in the land of the oppressor than his Irish heritage would have allowed him to admit – he spent most of his free time in a pub with Protestants who spent much of theirs singing about being up to their knees in his Fenian blood. We often focus on what we inherited from our ancestors, and these traits, beliefs and wounds form the spines of our own narratives, but what of all the other ways the story could have gone had we identified with or internalised different influences? I never regarded myself as being of Irish heritage, despite it clearly shaping my family. I never embraced Celtic Football Club – an emblem of Irish pride and resistance – despite its omnipresence in my childhood. Instead, class became the structuring force of my life, partly through material conditions of poverty, yes, but partly because I gravitated towards that defining theme for reasons unknown. That was the story I selected from the narrow range then available to me, but when you lay out all the elements that could have shaped my self-image – yet didn't – you begin to see how identity, beyond what we cannot control, is ultimately a construction. It's just another story. I could rewrite the whole thing right now, rearranging the pieces to suit whatever narrative or identity I prefer – that's how fickle identity can be. Given the aspects of my identity I've chosen to telegraph over the years, you might assume I come from a lineage of hardcore working-class militants. But you'd be wrong. The great irony of my family is that those who truly laboured – who left school early and went straight into a trade – were the least interested in class politics, or any politics for that matter. Only a few of us internalised the story of being working class, though I can't pinpoint exactly when or why I did. These choices, these inherited and rejected narratives, remind me that identity isn't a fixed truth, but a story we tell ourselves, one that can shift, evolve or even unravel entirely. And these questions may reveal a deeper truth. While it suits me to believe my character development followed a logical, linear progression – maintaining the face-saving continuity essential for driving the plot and keeping my story intact – I cannot say with certainty that any of it was deliberate. In some parallel universe, where my dad was the alcoholic parent and my mum the one who stuck around, I could just as easily have been a flute player in the Orange Lodge. The events of my life only became formative in hindsight, after I had already developed a vague sense that my identity needed a story to explain it. I'm not saying aspects of that story aren't true or that our identities are entirely fabricated. Nor am I suggesting we throw important parts of ourselves under the bus when the going gets tough. What I'm getting at – and what I now invite you to consider about your own narrative – is that while we like to believe we have made choices with intention, and that somewhere deep within us lies a fundamental, defining essence, the truth is that many of us simply don't know. Not if we really think about it. This is good news for those of us who have unconsciously adopted certain facets of identity out of fear or insecurity, stemming from traumatic wounds. We may experience a brief sense of relief when gaining the acceptance that we crave, or the sense of belonging we've lacked, but when that guitar is out of tune, even slightly, we can usually hear it – even if nobody else does. Am I defined by trauma and class? Or are these just the frames I've cobbled together in an attempt to impose order on the ceaseless maelstrom of a chaotic inner world. Is my identity fixed, or is it partly a story that can be redrafted? These questions don't have easy answers, but I believe they are worth asking in any discussion about releasing trauma. One long-term impact of untreated trauma is how it can fracture our sense of self, fragmenting our identity until it's unclear what's real and what's performed for survival. Some define themselves entirely by suffering, others adopt a hardened survivor persona, rejecting vulnerability. Either way, the past dictates the present. Emotional instability makes self-image unreliable – confidence one moment, doubt the next. Many seek meaning through work, activism or relationships, believing achievements or external battles will soothe their internal wounds. But without healing, these pursuits may become distractions rather than true expressions of who we are. Healing isn't about erasing the past or crafting a fixed identity; it's about learning to hold complexity, recognising how trauma plays tricks on us and resisting its attempts to dictate our existence. If we're performing identity to some extent – curating, adapting, revising – then the real challenge isn't simply defending the roles we've chosen, but asking ourselves whether we chose them at all. The stories we tell ourselves – flawed and exaggerated as they often are – help us make sense of life. But if we truly want to heal and grow, especially from trauma, we must be willing to examine those stories critically. We must confront the uncomfortable truths we have buried and reclaim the parts of ourselves we have abandoned in pursuit of safety. Just as an unjust society inflicts deep suffering by forcing people to hide who they truly are, using shame and bigotry as a weapon, we may also collaborate in our own quiet torture – choosing to suppress truths about our nature, beliefs or aspirations, to the detriment of our own wellbeing. What this all boils down to, my friends, is that clinging desperately to identities that no longer fit is bad for our health. We construct backstories that make our existence seem coherent, telegraphing only those parts of ourselves we find acceptable, hoping the communities we align with will embrace us. But what of the shameful, complicated, and unspoken truths about who we are? The things we wouldn't dare share publicly. Aren't those also part of our identity? Or should they remain hidden – until those dark nights of the soul, when the performance fades, and we are left with nothing but the raw, bleeding wounds we pray no one else will ever see?

'Many survivors have such profound deficiencies in self-protection that they can barely imagine themselves in a position of agency or choice.'

Judith Lewis Herman,
Trauma and Recovery (1992)

CHAPTER 13

When Identity and Injury Collide

What does over-identifying
with victimhood cost us?

Now, let's travel back in time again, a few years on from those days at my grandparents'. It's a dreary afternoon in the year 2000 and my sense of identity is about to shift – dramatically. I'm sitting on a stage in an assembly hall that doubles as a gym hall in a partially closed school in Glasgow. My classmates are playing badminton, each ten-minute game determining who moves up or down the league table. My best friend, Sammy, a hip-hop enthusiast, is at the top of the league. He's playing against a boy called Scott. But the game isn't going well – Scott keeps using his racquet to hit the shuttlecock off my back. He's been doing it for three minutes, trying to intimidate me, and it's working. The only reason I'm not playing is that I'm injured – sort of. I've shown up to school on crutches I borrowed from a friend's house. Truth is, I don't actually need them, but I've decided to fake an injury to get out of doing PE. A few days earlier, I was 'hacked' multiple times on a gravel football pitch by another classmate who whispered

'Fenian bastard' in my ear as he did it. Unable to take part in gym today, I've been instructed by a teacher to sit on the stage and complete some written work. Scott's behaviour makes it clear he assumes that since I didn't stand up for myself before, I won't today. I'm unsure why he's taken a sudden dislike to me – we've had no prior issues. I deliver the evening paper to his house. But that's just how this environment works: one day you're declared an enemy, and suddenly you're thrust into the game theory of survival – knowing you'll eventually have to fight back, even if you don't want to, to deter future conflicts. What neither Scott nor I understand yet is that a new, pervasive influence has entered my life. An artist I stumbled upon a year earlier, with baggy blue jeans, a white T-shirt and peroxide-blond hair, is profoundly reshaping my story. His music has become my lifeline, redrafting my narrative. For the past year, I've hung on his every word, absorbing them with a devotion that would put even my favourite teacher's best lessons to shame. Eminem, real name Marshall Bruce Mathers III, is fast becoming a global phenomenon. Before his songs were routinely beamed into homes worldwide, I got my hands on a cassette tape of *The Slim Shady LP*. Unlike most rap I'd heard – focused on street politics, poverty, police brutality, drugs and violence – Eminem's lyrics centred on his dysfunctional upbringing and his fraught relationship with his mother. His music struck a deep chord with me, fundamentally altering my trajectory. The media is awash with reports about how violent, misogynistic and homophobic his music is, yet he's the only person I've seen on television, in my entire life, who I feel in my heart is speaking directly to me. And what he's telling me is that bullies understand one language and one language alone. At the end of the school day, when the bell goes, I drop the crutches and land as many blows on Scott's face as I can, taking him completely by surprise. That he wasn't expecting me to erupt in anger is precisely what gives me an advantage. Had I gone along with his suggestion, that we meet off the school grounds to dook it out, I'd have put myself in danger – it's unwise to get the better of someone who thinks they have less to lose than you do. My strategy is not to humiliate or hurt him, it's simply to fire a warning shot to Scott and everyone else that I'm sick and tired of being taken for a fucking mug – and it works. He barely gets a hit in before the teacher enters the dressing room to break up the fight. Everything goes according to my plan, though I am immediately suspended from school. Scott never bothers me again.

Fast forward three years. School is a distant memory. I'm living in supported accommodation for homeless and estranged young people. My wardrobe consists of baggy second-hand clothes, trainers with fat laces and baseball caps or woollen hats. I've even adopted a new name: Loki. Gone are the days of avoiding conflict. Now, I actively seek it. I'm no longer the anxious boy hoping not to draw stares. I'm the one glaring across the room. Barely 20 years old, I'm convinced I'm the best rapper in the country and determined to prove it. All I care about is writing songs, recording and performing. I have dreams of becoming famous, making my family proud and the naysayers sick to their stomachs. In reality, my life is a shambles. I'm on benefits, have no real employment or educational prospects, and I'm drinking heavily and experimenting with drugs. Yet the gravity of my situation is lost on me. I'm wrapped up in a delusion about who I am, where I am, and how I got here. Estranged from my family, I've convinced myself I've been outcast and abandoned. This narrative bleeds into every song I write. The angrier the music, the more people seem to love it. In this new identity, I feel free from the awkward, misunderstood boy I once was. The acne is gone, the red hair is covered by a baseball cap and the girls seem quite happy to talk to me now. In the local Glasgow hip-hop scene, I've found a tribe. But unlike many peers who see hip-hop as a unifying force, I view it through the lens of competition and retribution. I'm fuelled by the idea of settling scores, too distrustful to take others at face value. I'm blind to my own contradictions, believing myself insightful while acting arrogantly and confrontationally. Words have become more than a means of self-expression; they are sharpened implements I use to perform surgical procedures publicly on anyone who crosses me. My story is now about how I was let down, how I am poor, how I am 'fucked up' mentally. Some of this is true, of course, but in leaning so hard into the victim facet of my identity, fortified as it is by anger and resentment, I overlook many of the privileges I enjoyed as a child. The sacrifices my father made to raise three of us as a single parent, and his constant encouragement to pursue my dream of a career in the arts despite immense social and economic pressure to revise my aspirations down and look for a traditional job. I was blessed with wonderfully skilled teachers at school, who never let me forget my potential – even if it meant pulling me in for a dressing down now and then. The joyous summers where days seemed to last for weeks. The aunts and uncles who went out of their way to spend time making those lasting memories you carry with you for life. The patience and loyalty of friends and their families who had always been there for me, who I mistreated or was mean to, or simply forgot all about when I assumed my new identity. My story was always about so much more than poverty and alcoholism, but somewhere along the way, whether through unconscious editing, or public fascination with working-class grit and hardship, the true richness of my childhood was lost in the retelling.

When Loki came into my life, it changed how I felt about myself and how I behaved. Some of these changes were positive – greater confidence, purpose and a sense of belonging – but some were for the worst – aggressive, distrustful, arrogant. I subconsciously selected certain traits to emphasise and certain traits to downplay. This was a full embrace of the toxic traits I'd done so well to resist as a teenager in Pollok and to fully inhabit this character, I would abandon other parts of myself – truer parts. The fact I always feared physical fights as a kid and would freeze as often as I would hit back, was a truth that had to be concealed. The fact I was scared of the dark when home alone, due to sleep paralysis I would experience as a result of exhaustion from drinking for days on end, was not something I was singing from the rooftops. And I wasn't the only one leaning hard on the fiction genre either. When I travelled to the Canary Islands to visit my then girlfriend, who worked there as a kids' entertainer, I arrived to discover she had made up a story about how I was signed to a record label – a fabrication I doubt her coworkers even believed. Perhaps she was embarrassed that I was actually living in homeless accommodation and on benefits, and ashamed that I was trying to make it as a rapper. At least, this is what I assumed given she felt the need to furnish her colleagues with a tall tale that reflected more generously on her decision to go out with me. Rather than see this as the massive red flag it really was, I happily obliged. Strangely, I enjoyed playing a part for her and in some ways began believing the lie myself. I was comforted by the idea of being someone she found acceptable and the prospect of not having to work so hard for her love and affection. I sought refuge in my identity as Loki – a victim of circumstance – and put out of mind how my wounds showed up in our relationship as an overbearingly anxious attachment style, in constant need of reassurance. While I was caught up in this turbulent teenage romance, the thought occurred that I might reconnect with family and ask for help, but I was developing a sense of comfort in isolation – a sad fact I rebranded as a fierce independent streak and folded into my identity. Returning home would only create dissonance within me. Unbeknownst to me, my long absence left my siblings and other relatives I was close to, like cousins, little choice but to create their own stories to explain why I wasn't there – or to account for the inebriated version of me that made rare appearances now and then. Unsurprisingly, these stories reflected less well on me than the one I'd created about myself. In their stories, I was the junky and the abandoner. In their narratives, I was the self-centred arsehole who didn't care. They were the victims, not me, though it will come as no shock to learn I didn't see it that way at the time.

*

There is no greater sin in the world of trauma than that of dismissing, invalidating or questioning victims. There are good reasons for this. It takes immense courage to come forward and disclose pain. In a therapeutic context, affirmation of victimhood builds trust, allowing someone who has been wronged or traumatised to open up and begin organising their experiences into meaning. In a technical and legal sense, a victim is simply someone who's experienced a bad thing that wasn't their fault. Today, the term 'victimhood' is almost as contentious as 'trauma'; it means something different depending on who you speak to. For some, to claim victimhood is an act of empowerment. For others, it's a mentality that may hold you back. How the term is applied often depends on the prior experiences, biases, loyalties and beliefs of the observer. Victimhood is a deeply layered and emotionally charged concept, especially when viewed through the lens of trauma. At its core, victimhood acknowledges the experience of harm or injustice. It serves an important function – validating pain, offering a framework for understanding what has been endured, acting as a rallying point for seeking justice or healing. For many, claiming victimhood can be the first step towards reclaiming agency, as it provides a way to contextualise suffering and seek solidarity with others who share similar experiences. But when victimhood becomes intertwined with identity, it can be both a source of strength and a trap, obscuring the ways unresolved trauma may ripple outwards and affect those around us. Trauma shapes how we see ourselves and the world. If we have endured profound harm, recognising we are victims can feel like a lifeline – a way of making sense of chaos or cruelty. There's immense power in saying, 'This happened to me, and it wasn't my fault.' It provides clarity, a way to anchor our experience, and perhaps even a call for recognition. For some, speaking out is often the only means of reclaiming a narrative in which a perpetrator's abuse or backstory has been wrongly centred over a victim's. Yet, as we've explored, trauma doesn't remain static. When unresolved, it can manifest in unconscious patterns – emotional withdrawal, heightened defensiveness and even the reenactment of harm in relationships. These behaviours, though protective in intent, can sow confusion or pain in the lives of those close to us, creating cycles that we may not even realise we are perpetuating. In some cases, we may become so attached to the notion of our victimhood that we develop a particular mindset which frames subsequent experiences and relationships. In psychology, this is referred to as a victim mentality. A victim mentality is a learned way of thinking (a response or adaptation) where a person consistently sees themselves as powerless in the face of adversity. It often develops from real experiences of suffering or injustice, but over time it becomes a habitual lens through which challenges are interpreted. It's a mentality I see in many addicts before they finally accept they have a problem and decide to get sober – a mentality I developed in the throes of my own alcoholism. People with a victim mentality may externalise blame, struggle to take responsibility for their role in situations and expect negative outcomes as a default. However, this mindset can be changed with self-awareness, support and a shift in perspective; a delicate process, especially when that person has endured real harm to the extent the wound has tipped their nervous system into permanent vigilance. For those who develop a victim mindset, their victimhood is a story they tell themselves subconsciously to make sense of the bodily condition of trauma and the increasingly precarious nature of their daily lives. It's essential to acknowledge that few people choose to view their lives through a victim frame; it's a trauma response for many people, and compassion must be extended. When living with trauma, life comes at us faster than we are able to anticipate and throws us off balance in myriad ways; for some, the result is a skewed perception of their own vulnerability, agency and culpability. In some cases, however, a victim complex (more severe than a victim mentality) may develop. A victim complex is more deep-seated and pathological. It is a psychological condition where a person becomes convinced that they are being perpetually wronged, regardless of evidence to the contrary. Unlike a victim mentality, which is a cognitive pattern, a victim complex becomes an ingrained identity. Those with a victim complex may reject solutions, reinterpret neutral or positive situations as threats, and resist accountability entirely. Their sense of victimhood fuses with their self-concept, often making relationships, work and personal growth difficult. A victim mentality is a mindset that can be unlearned, while a victim complex is an entrenched identity that actively resists change.

With that distinction in mind, let me now attempt to explore this sensitive theme of victimhood further, parsing out the complexities in the context of trauma and healing. If you feel in any way distressed at the prospect of this line of enquiry, I welcome you to take a break or skip to the next chapter – return when you feel ready. If, however, you feel safe enough to continue, please be reassured that I have taken considerable care in how I approach this delicate topic. While certain points may raise an eyebrow, or even a provoke a twinge of anger, this book was written with the singular intention of supporting victims and survivors in their recovery. It is my experience that at some stage in our healing journey, whatever has occurred, difficult questions must eventually be broached. Given the vast distance we have hitherto covered, now feels like an appropriate moment to pose some of the hardest we are likely to face. When toying with the idea of writing this book, the moment inevitably arrived when I was moved into action. It all began in 2023 with a now-deleted tweet by a mental health professional, known online as The Wounded Healer. The tweet, paraphrased, stated: The best way to test a person's character is to see how they treat you when you're at your most vulnerable. At first glance, this sentiment seems uncontroversial. Vulnerability – whether it stems from physical, emotional or economic challenges – does indeed create opportunities for exploitation by those of low character. Yet the statement struck a nerve with me. At the time, I was trying to support many people who'd fallen on hard times as a result of the Covid pandemic and the economic and political chaos that ensued. A spate of local suicides over the previous years, which included my friend and musical collaborator in 2019, left me on edge and fearing for the lives of people I cared about who I knew were struggling. Of course, the reality of supporting people with mental health challenges is a world away from the platitude-ridden social media landscape where, apparently, all that's required is an empathetic affirm-only approach – in many cases nothing could be further from the truth. Anyone who's walked alongside a loved one in a genuine mental health crisis will tell you that often the biggest barrier to supporting them is their belief that they don't need help – and that anyone trying to help is the real problem. Even if they do vaguely sense they are unwell, they will resist any course of action you suggest. When they do finally accept the gravity of their situation, that's when the obstacle course of public services begins and it's a race against the clock to align professional support with your friend or loved one's desire to engage with it. When I read the tweet in question, it felt like a slap in the face to be honest. I'd given countless hours to people in crisis, gone through doors expecting to find bodies hanging from ceilings, counselled broken families on getting their broken loved ones into rehab, and attended various funerals, consoling dozens of grieving friends and relatives. With no medical training or expertise, I'd assigned myself the role of first responder to many in crisis and was beginning to feel burned out. I even recall contacting the Samaritans for advice about how best to approach people who were clearly suicidal – I was regularly waking up to audio suicide notes or drunken diatribes and didn't want to make it worse by misspeaking – but within five minutes of the call, the responder asked simply: 'And who is supporting you, Darren?' An incisive question, to be fair. In truth, I was having my character judged, often by the very people I was trying to support, whose mental health and addiction issues became so acute that they had begun losing all grip on reality. These relationships had become a one-way street where I gave and gave and gave and received very little in return but stubbornness and disrespect. I imagined The Wounded Healer's tweet being read and internalised by these people, or cited as evidence that I was not in fact helping them but abusing them. People who were deeply and undeniably vulnerable and also creating great distress in the lives of those around them – including mine – while believing themselves purely victims. The Wounded Healer has built a significant following as a public figure in the mental health space. Known for blending professional expertise with personal anecdotes, he often discusses the complexities of healing, accountability and interpersonal relationships. His insights resonate widely, offering comfort to many navigating mental health struggles. I enjoy much of his content and my critique here is not an attempt to cast doubt on his entire body of work nor his therapeutic approach or character. However, like many who tread the line between professional advice and social media engagement, some of his statements, such as the tweet in question, have sparked debate. As we've covered at length in this book, I've faced vulnerability in many forms – battling alcoholism, drug addiction and the accompanying mental health struggles. In those moments, however, I wasn't just vulnerable; I was also difficult to deal with. I resisted guidance, dismissed advice and perpetuated cycles of harm in my life. While my socioeconomic conditions and mental health played a significant part in prolonging my struggles, as did my untreated trauma, I cannot ignore the accountability I bear for rejecting difficult truths presented to me along the way. The tweet, though well-meaning, felt reductive. It encapsulates a popular social media tendency to oversimplify complex issues, particularly around mental health. In this case, the notion of victimhood. While it may offer validation or comfort to some, for others – like me – it risks reinforcing unhelpful narratives about the part we, the wounded, may play in our continuing suffering. The crux of my discomfort lies in the implied expectations embedded within the tweet. When we're vulnerable, should we really be encouraged to assess the character of those around us? It's reasonable to hope we won't be ridiculed, dismissed or invalidated, of course. Yet vulnerability – particularly mental health-related vulnerability – can distort our perceptions of others' intentions. Encouraging someone whose ability to assess their true circumstances may be compromised to pass confident judgement on others' character is fraught with potential for misinterpretation and harm. Moreover, the tweet overlooks the complexity of human relationships and the limitations of those around us. It assumes others possess the energy, knowledge and resources to consistently meet our needs but for reasons unknown are refusing to do so. When they don't or can't show up in the ways we want or need, are they inherently of low character? Or are they, too, navigating their own vulnerabilities and limitations? What if our assessment of our own needs is wide of the mark? What if the crux of our problem is centred in the fact that we have lost all sense of our true needs? How many addicts beg loved ones for money for drugs, to meet their need for a fix? How many people refuse social contact, feeling it safer to remain alone, when chronic loneliness is often as bad for your mental and physical health as smoking 20 cigarettes a day? In my own case, there have been countless times when what I thought I needed was the least of my real priorities. During my struggles, I often viewed relationships through a lens of expectation – expecting others to meet my needs without fully considering their circumstances. This mindset, fuelled by my addiction and mental health challenges, led to feelings of betrayal and resentment when people withdrew or couldn't provide the support I sought. If ever challenged, I would feign offence and pull away, seeking refuge in a victim mentality, so as to avoid difficult conversations where I may have to sit in the truth for a time. In hindsight, I realise the unfairness of these expectations. In an ideal world, we'd all know what we need, be able to articulate those needs and see them met when expressed. But this world is far from ideal. Others have their own struggles, and their inability to meet my needs doesn't necessarily reflect their character. Prolonged vulnerability such as that which may develop in those with untreated trauma can create a dangerous paradox. On the one hand, people must approach us with sensitivity, understanding the fragility of our state. On the other, true support often involves challenging us in uncomfortable but necessary ways. Without this, we risk stagnating, shielded from accountability by our perceived vulnerability. I've experienced this first hand. During my addiction, I surrounded myself with people who to me felt safe because they wouldn't challenge me – those who either shared my blind spots or feared upsetting me. This 'cosy consensus' allowed me to avoid confronting uncomfortable truths, perpetuating my destructive behaviours, which were often validated and cosigned rather than called out. True support, however, is when compassion is balanced with honesty, where validation doesn't come at the expense of having our skewed perceptions challenged when necessary. In my recovery community, I am known for my willingness to risk offending others by speaking a hard truth or two when necessary. I never do so thoughtlessly; it's often a high-risk strategy, adopted when all others have been exhausted. When someone becomes so vulnerable that their life may be in danger through overdose or suicide, and the only action they seem willing to take is retelling the story of why their lives are so unbearable, sometimes you are called upon to take a more radical approach. Rather than endlessly affirm their interpretation of their circumstances, you must present them with a stark choice: either you take action to move through these difficulties, with my full support, or you descend deeper into your resentment, selfishness, fear and dishonesty. I don't take this approach lightly, but rarely does it fail when delivered with love and compassion. The Wounded Healer's remark, in all its well-meaning simplicity, also fails to account for the broader societal and communal context in which so much suffering occurs. In communities grappling with widespread mental health struggles, addiction and socioeconomic hardship, everyone is vulnerable to some degree. From which reservoir, exactly, can this undiluted, perfectly pitched form of support be drawn? I deal with people who've been chewed up and spat out by life, ingested by the criminal justice system and shat out onto the streets. To reach and build a rapport with people in this mould, who already assume everyone is of low character due to psychological projections and trauma, you require more than a drive-by Twitter thread. And some days, success or failure depends on what you, the person reaching out, has going on in your own life. Some days, I just don't have the wherewithal, the time or even the patience to show up in all the ways someone else might need. Still, I have to show up. The collective strain – exacerbated by crises like lockdowns, austerity and the cost-of-living crisis – means many people lack the capacity to meet others' needs, no matter how much they might want to. This is particularly true in my community, where the ripple effects of suicide, addiction and other deaths of despair weigh heavily. People aren't just bearing wounds from childhood, they bear scars from these untreated wounds being reopened by systemic breakdowns that reverberate around their lives and the lives of those within their support networks. Each loss amplifies the burden on those left behind, creating a frayed network of individuals trying their best to support one another while barely coping with their own challenges. In such conditions, judging others' character feels deeply unfair and unproductive. When victimhood becomes a central aspect of identity, it can blur the line between understanding one's pain and perpetuating harmful patterns unintentionally. Trauma often creates a dual reality: on the one hand, we who have suffered are undeniably deserving of compassion and understanding; on the other, our pain doesn't absolve us from responsibility for how we show up in the world. Someone who endured neglect as a child might become overly controlling or emotionally unavailable as a parent, inadvertently passing down wounds they sought to escape. A friend consumed by their own grief might unintentionally overshadow others' struggles. These dynamics highlight the paradox of victimhood: it demands to be acknowledged, but when it dominates identity, it can limit the survivor's capacity to see beyond their pain and connect meaningfully with others – a profound barrier to healing. Disagreements may be perceived as attacks, mistakes interpreted as betrayals, and others' pain minimised in the face of our own. To examine these dynamics can feel like an invalidation of our suffering, but avoiding this reflection risks locking us in a cycle where the need for validation outweighs the possibility of growth, leaving harm to reverberate unchecked through our relationships and communities. True healing lies in holding space for both truths: that the wounds of the past are valid and deserving of care, and that healing also requires reckoning with how those wounds might cast shadows over others. What was the reality of my identity back in my early adulthood, when I viewed life through a lens of victimhood? Beyond my self-image and those immutable characteristics, who was I really? I wasn't a bad person, but I wasn't quite the victim I believed myself to be at the time. Are we whoever we think or claim we are, or are we also partly defined by what we do? Is it authentic to assert certain facets of my identity – my trauma, my working classness, my victimhood – while downplaying or dismissing the impacts of my wounds on others? What about the money I stole from my grandad because having cash in my pocket gave me more confidence? Shouldn't I integrate thievery into the story? Or items I pocketed from house parties because I thought the hosts were affluent snobs – doesn't this make me as entitled as I believed they were? What about the loving and compassionate girl-friend I met years later, who held space for my many endless needs. The one who I came to find boring because she wasn't toxic to my nervous system and who I drunkenly left with no explanation. Where's the keynote about all that? The friends I lied to or manipulated for drugs. The dying family members whose medications I helped myself to, feeling entitled as they would only get more prescribed anyway. And then my granny, who I also stole drugs from, who in many ways raised me as a son. She would constantly contact me when I was wayward, asking me how I was, inviting me down for visits, but I grew irritated by her. She was cramping my style. I took her presence in my life for granted and it's something I deeply regret to this day. When she died suddenly from a hospital infection, I made it all about me, too. I didn't visit her on her deathbed because I didn't want to stop drinking that day and going to the hospital would have meant being around family who'd see me for the alcoholic mess I really was. Yes, I was a victim of certain circumstances – poverty, dysfunction, class inequality – but that's not all I was. Untreated, my wound mutated into something else. Obeying the commands of my whispering demons, with no plans to seriously tend to those wounds despite the vast amount of support made available to me, I became selfish and self-centred to the core; my personality a conflagration of trauma responses directing me to self-soothe at the expense of everyone around me. I wasn't paying my bills or debts, I wasn't checking in with my siblings, and I held little space for the wounds of others unless there was something in it for me. I had wonderful, at times persuasive, impassioned justifications for all of this of course, but in the cold light of day, there came a time when I had to start 'believing' and 'affirming' the many victims of *my* selfish behaviour. My mission back then was holding everyone and everything accountable but myself. The integration of victimhood into my identity was both a shield and a mirror. It protected and affirmed me but also reflected the ways pain shaped my inter-actions with the world. My trauma demanded recognition, but it also required a reckoning. Moving beyond the confines of victimhood – not by denying it, but by allowing space for com-plexity – is absolutely vital, though certainly not easy. I speak here not from some remote, privileged vantage point that knows nothing of struggle, but from genuine experience of adversity. Whatever I suffered, whatever my pain, I eventually developed a victim mentality as a response. As a result, I often could not see where I was at fault, how I was prolonging my own suffering, or how my urge to self-soothe and protect was harming people I loved. Many of us love proclaiming how 'hurt people hurt people' without truly considering this well-worn phrase's implications. Many of us rightly take refuge in the status granted to us as victims who carry trauma but are less likely to embrace the thornier truths that often come with this special designation. I was a victim of many harms, but acting from woundedness, I also harmed others. In my story, I was the unlikely hero who made it against the odds, but in the narratives of others caught helplessly in the crossfire of my internal war, I was most certainly a villain. I do not sit in harsh judgement of myself, nor do I expect anyone else grappling with these complicated questions to either, but that's what trauma did to me. That's what trauma does to many of us. And that's the side of the story about trauma nobody wants to tell.

'Trauma in a person decontextualized over time can look like personality. Trauma in a family decontextualized over time can look like family traits. Trauma in a people decontextualized over time can look like culture.'

Resmaa Menakem,
My Grandmother's Hands (2017)

The Politics of Trauma

How does widespread trauma
reflect broader societal injustices and who
is ultimately responsible for healing?

Trauma doesn't happen in a vacuum – this much we understand. It unfolds within the fabric of relationships – relationships that were once (or ought to have been) the source of our security, belonging and support. At its heart, trauma is a violation of trust in these connections. When that stability is shattered – whether by betrayal, abuse, neglect or calamity – the wounds cut deep. These relationships extend beyond personal connections and include our attachments to communities, employers and the institutions that shape and govern our existence. Even our relationship with the natural world relies on a fundamental trust – that every storm will eventually pass, and that the ground beneath our feet won't give way. Consider survivors of natural disasters or catastrophic accidents. Think about victims of abuse betrayed by the very criminal justice system meant to protect them, or a child excluded from school – the only safe place in their life – for 'challenging' behaviour. Though these scenarios seem disparate, they share a common thread: the wounds inflicted are deepened by the shattering of faith in a safe and stable reality. When you step onto a train, you trust the brakes will work. When you board a plane, you assume its navigation systems are sound. When you lay a towel on a sunny beach, you do so believing there's no tsunami racing towards you, unseen. Trauma is the moment that trust is obliterated – when you realise you are not OK, and that realisation is so sudden and overwhelming it exceeds your capacity to cope. In many cases, trauma doesn't just arise from the initial event but is aggravated by the systems we've created – systems ill-equipped to handle trauma's effects. These structures, far from offering support or healing, often complicate matters for the vulnerable. Take, for example, the children present during a dawn raid on a suspected drug dealer's home. Police officers don't pause to consider the impact of their violent entrance – shouting, smashing doors, brandishing weapons – on the young minds witnessing the chaos. The children are rarely suspects, but they become collateral damage. Consider the way social security interacts with benefit claimants or those with physical or mental disabilities. Communication from these agencies is often cold, bureaucratic and impersonal, designed to elicit compliance rather than understanding. For vulnerable people with trauma living in their bones, these interactions feel less like assistance and more like psychological waterboarding – a barrage of letters, forms and demands that provoke panic and dread. Even healthcare systems, which should be sanctuaries of compassion, may unintentionally create or exacerbate trauma. Children undergoing radiation therapy for cancer are routinely separated from their parents, left to face the frightening experience alone. Mothers struggling with drug addiction – often a symptom of unaddressed trauma – are imprisoned for

petty crimes, further fracturing families, thus imbedding generational cycles of dysfunction and despair. Young offenders,
many themselves victims of abuse and neglect, are locked in
cells for 23 hours a day, isolated in facilities teeming with other
violent offenders. These practices don't just fail to address
the root causes of their behaviour; they aggravate the trauma
that drove it in the first place. Trauma and social inequalities are deeply intertwined, a relationship long recognised in
trauma theory and advocacy. Judith Herman, a pioneering
figure in the study of trauma, notes in *Trauma and Recovery*:
'The conflict between the will to deny horrible events and the
will to proclaim them aloud is the central dialectic of psychological trauma.' This duality not only underscores the
personal struggles of survivors but also the broader societal
forces that perpetuate cycles of harm. Herman further emphasises that 'recovery can take place only within the context of
relationships; it cannot occur in isolation'. Healing therefore
requires not only personal effort but also the dismantling of
systemic inequities. Efforts to reduce economic inequality and
expand access to education, as well as wider support services,
are crucial in creating environments where survivors can genuinely thrive. These impassioned calls for change reflect a belief
in the interconnected nature of individual and collective healing. Critiques of systemic inequality through the lens of trauma
range from moderate to radical. Some call for a liberal, rights-
based approach which argues for reform but stops short of
recognising the role capitalism clearly plays in creating the
relational and material scarcity in which trauma often thrives.
For some however, this does not go far enough; every problem
we could conceivably experience is attributed to free-market
ideology. Many of trauma discourse's opinion-shapers go

further than offering mere observations of trauma's systemic roots; politics becomes central to the analysis, and therapists and survivors are encouraged to view inflamed nervous systems not as injuries to be regulated or soothed but as natural responses to injustice that should form the basis of revolutionary action. Thinkers in this radical mould suggest that creating a fairer society is not simply an adjunct to healing but a fundamental part of it. That there can be no healing without vast and sweeping social change in which survivors must actively participate. Trauma work, when approached in this way, becomes not just an act of individual survival but also a call to actively reshape society. Only then, argue many radicals, may we hope to create a world where healing is not a privilege but a right, accessible to all. This, my friends, is where I part ways with many on the left flank of our ongoing dialogue about trauma. For me, this laudable idealism centres political ideology rather than healing. In theory, of course, I understand the argument completely and once held the same convictions. In practice, drawing from many years of experience as a social and political campaigner, I'm quite unsure how you can guarantee a right to freedom from trauma in a world where we've yet to crack the political conundrum of ensuring everyone on the planet has access to clean water and food. But this more politicised view of trauma does pose some interesting questions worthy of interrogation. What might freedom from trauma look like. Is it achieved by simply providing rights, resources and facilities for those in need? Or is true and complete freedom achieved when citizens possess the means to provide this for themselves, irrespective of the will or competence of decision-makers? And who is ultimately responsible for creating the conditions for this empowerment to occur?

Perhaps my own experience of breaking free from the chains of addiction may shed further light on my perspective.

In January 2013, at the end of a seven-day drinking spree, about to open another bottle that would tide me over until the off-licence opened, a woman named Katie whom I have never met sent me a message on social media. It was a link to an article that would snap me out of my drunken daydream: '6 Harsh Truths That Will Make You a Better Person'. I figured it would pass some time and began reading. By the time I got to number one on the list, 'Everything Inside You Will Fight Improvement', the urge to continue drinking quite simply left me. What had occurred was the profound psychic shift only a high, targeted dose of the truth can bring about. For years, I had dined out on my trauma, my losses, my grief and my anger, using them as excuses of varying plausibility to justify my descent into alcoholism. It's true that I faced significant adversities in my youth – they had a lasting impact on my character and emotional nature, for better and for worse. But at some point, I lost touch with the idea that a better life was even available to me. I became resigned to the misery of depression, the painful solitude of self-isolation and the invigorating, if toxic, effects of my righteous anger. I was sick because the world was sick. And I couldn't get better until the world improved. My solution was to try and make the world a better place from within the confines of my sick mind, but oddly, nothing ever seemed to work. The article read like a mug of cold water thrown in my face. My blunted faculties sharpened. A self-awareness pierced the thick fog of denial. And after I finished the piece, I poured what remained of my alcohol down the kitchen sink and told my long-suffering flatmate I was done with the drinking. From that day, my recovery began in earnest, and I wouldn't lift a drink for a further two years. Save for a few slips along the way, I have been alcohol- and drug-free for most of the last 12 years. How was this achieved? Where did the power to stop drinking come from? A power that had eluded me almost every day of my twenties. Did it come from the state? Was it supplied by the market? I got sober in run-down community centres and churches, where no experts or professionals were present. Indeed, my many inter-actions with public services throughout the years played some part in my adopting the false belief that I would never get free of the addiction. Instead, I got well in rooms where the advice dispensed came from other sufferers of the problem, not from medical professionals, activists or theorists. Mutual-aid groups where there are no hierarchies, no professional titles and no state or private funding. I learned to traverse the greatest chal-lenge I have ever faced as an individual – the illness of addiction – merely by following the suggestions of those who had gone before me. I did this in a community where it is understood that we could only ever hope to be of any meaningful or last-ing use to our community by first making ourselves accountable. Accountable for whatever part we play in our adverse circum-stances, accountable for the harms we have caused, for our dishonesties, our attitudes and our behaviours, and commit-ting to living by certain principles. In those rooms, we achieve freedom from addiction by learning to reconcile our individual needs and desires with the world as it is – not as we need or would like it to be. We claim our right to liberty from addic-tion by taking that freedom ourselves. I am well aware that there are those whose challenges are too profound to eschew medical expertise, whose economic adversity is too acute to simply think your way out of, or who face persecution or oppression that only political action and resistance will remedy. But in my experience, there are lessons as to how anyone, within reason, might still better orientate themselves in the face of a problem rooted in systemic inequality, so as to lighten the individual burden and make themselves more useful in any wider struggle. It's in this space that the two distinct ideological tracts of the Trauma Industrial Complex – advocates who believe in systemic change through collective action and those who view the individual as the bulwark of a free and fair society – must be viewed. The ideological tug-of-war at the heart of debate about trauma is merely an echo of a much deeper struggle between competing visions of the Western world. It's also a struggle that has, in many ways, defined my own politics. Two powerful stories, seemingly impossible to square off, spooling endlessly alongside one another. On one side, the ideology of individualism, with its emphasis on personal responsibility, self-reliance and achievement, has profoundly shaped how we talk about trauma. This framework often detaches adversity from the social conditions in which it arises. Trauma, in this context, becomes a personal struggle – a wound to be healed through introspection, therapy or self-help. Recovery becomes an internal project, requiring resilience and effort on the part of the individual, with little attention paid to the broader systemic forces that created the harm in the first place. Survivors are encouraged to focus on their own healing, as though their suffering exists in a vacuum, unconnected to structural inequalities. This decontextualisation of trauma has profound implications. When trauma is reduced to a personal problem, it risks pathologising the individual while ignoring the systems that perpetuate harm. This individualisation is further entrenched by the commodification of healing. The burgeoning wellness and self-help industries have turned trauma recovery into a lucrative market, offering courses, apps, retreats and other individualised solutions that, while sometimes helpful, rarely challenge the societal systems that produce trauma on such a scale. The individual, not the system, becomes the locus of change, leaving the status quo unchallenged. In this light, the decontextualisation of trauma can be seen as a reflection of a society uncomfortable with confronting its own complicity in harm – a society telling itself an incomplete story. It is far easier to ask individuals to carry the burden of recovery than to undertake the collective responsibility of addressing the inequities and injustices that underpin so much human suffering. On the left, where most trauma advocates fighting for systemic change reside, these individualist delusions and refrains are robustly challenged. The ludicrous notion that one person, by sheer force of will, can offset multiple deprivations merely by changing their attitude or by working harder invites rebuke. In recent years, trauma advocacy has increasingly intersected with political movements, transforming the work of healing into a dual mission: personal recovery and societal change. Healing is no longer seen solely as an individual endeavour but as a collective responsibility to foster a more compassionate and equitable world. Those of a more socialist disposition, who believe the plight of the poor is best served through collective action and the state, are, in my view, correct in their analysis. But this perspective almost always lacks any analysis of the role that the individual may still play in producing better circumstances – even within that unjust context. Furthermore, on the left, it can be offensive to suggest that not all individual problems are rooted in failing systems or solved by allocation of greater resources. Some of our problems are of our own making and therefore our own to solve. While we are rarely responsible for the traumatic wounds we bear, healing from trauma requires countless acts of individual exertion. While relationships and support are essential components of recovery, survivors will be called upon to lead their own healing – irrespective of the circumstances in which they were wounded. This simple truth, however, is deeply provocative. Despite my lived experience, I have been accused of engaging in what is tantamount to victim-blaming for pointing out the various ways an individual who lacks insight, self-awareness, or like me becomes gripped by the malady of addiction, can make their experience of a problem rooted in structural inequality (like trauma or addiction) much harder than it has to be. And I have been criticised for asserting that in many cases, before a person can begin to consider acting meaningfully upon an unjust society, they must also become willing to change themselves – the most radical thing some people will ever do. The need for profound change is undeniable. The question therefore arises, should we grant the ascendant notion that the wounded themselves must rise up and reorder the world around them through political action: what would that look like in practice? What obstacles might they face both internally and externally? Are comrades identified purely on the basis of political affiliation or is greater discernment necessary on the part of would-be changemakers, still tending to their wounds, who may be vulnerable to political chicanery masquerading as benevolent allyship? And how might untreated trauma predispose activists to further harm in the twenty-first century, where the very technology essential to participate politically is optimised to prey on our fears, resentments and vulnerabilities? As we have explored in great depth so far, too often, those carrying unhealed wounds are thrust into the forefront of the political crusade. Storytellers are burdened with the emotional labour of sharing their narratives, only for their stories to be appropriated by facilitators. These narratives, used to raise awareness and drive reform, usually take two pathways: radical campaigns demanding swift, systemic change and more pragmatic efforts pushing for incremental progress. Understanding their dynamics helps clarify the tensions beneath the rhetoric. Radicals challenge entrenched power structures, using lived experience to create urgency and spark heated public debate. Calls to defund the police or introduce juryless trials for sexual offences exemplify this approach – bold, uncompromising, but often detached from political realities. While radical campaigns attract media attention and shape public discussion, they frequently collapse under their own weight, with failures absorbed into a mythology of resistance rather than prompting strategic reassessment. Criticism is framed as dismissal or invalidation, fostering a siege mentality among activists. When radical proposals gain traction, compromise becomes inevitable and experienced campaigners adjust accordingly. Less seasoned radicals, however, cling rigidly to their positions, prioritising ideological purity over practical outcomes. For them, being seen to concede ground risks alienation from their tribe – an identity rupture too severe to bear. This rigidity is often less a mark of conviction than a symptom of political inexperience, emotional immaturity or misplaced tribal loyalty. In such spaces, personal experience and politics blur; the personal becomes political, with no boundary between the two. The charge is led by emotion, rendering individuals, campaigns and even entire movements dysregulated from the outset.

Many of the most dedicated and intense campaigners are themselves survivors of trauma, which grants them deep insight but may also make them notoriously difficult to engage with. We buy wholesale into the idea that as people who have direct experience of an issue, we are the sole authorities. The only out-group opinion we're interested in hearing is that which validates our own. Some of the most effective activists – tireless, fearless and unwavering in their pursuit of justice – are also among the most unhappy, dysfunctional and wounded figures in the political sphere. Untreated trauma renders us inefficient politically and highly vulnerable in our personal lives. Our pain fuels extraordinary efforts, and when we succeed we make a real difference. Yet for some, activism itself becomes a form of dissociation. The political struggle, no matter how futile or miscalculated, feels safer than the uncertain, personal work of self-examination. For too many, introspection is an indulgence, a sign of weakness. But this story we tell ourselves – like all stories – can be a way of keeping deeper truths at bay. For many, political campaigning is simply a form of emotional avoidance; political issues become the means by which failure to confront ourselves, and the role we as individuals may play in some of our suffering, is justified. Unsurprisingly, however, activists in this mould rarely see it that way. Pragmatists, on the other hand, present a stark contrast to radicals. Cooler in temperament, they are often dismissed by radicals as lacking urgency, conviction or emotional investment. But their approach differs by design. For pragmatists, reform is a marathon, not a sprint. They operate within the civic sphere, away from the glare of public scrutiny, preferring slow, bureaucratic negotiation over high-profile and bruising confrontation. This behind-the-scenes work, largely

invisible to the public, occasionally surfaces through carefully managed press releases, publicity stunts or public events,
but real business is conducted between advocacy groups and
decision-makers. A sprinkle of lived experience aids the optics
but is often tokenistic. Unlike radicals, pragmatists accept the
constraints of the system, pushing for reforms that are often
incremental, politically safer and stripped of socioeconomic
context. Many have worked within the very institutions they
now seek to change – former police officers advocating for drug
reform, ex-third-sector executives reviewing the accountability
of charities, or former politicians critiquing democratic institutions. However, this proximity to power is both a strength
and a liability. Pragmatists lack mechanisms to hold decision-
makers accountable when promises are watered down or fail
to materialise. Their reliance on institutional relationships
– for funding, influence or a seat at the table – leads to unspoken compromises. What passes for 'grown-up politics' often
means looking the other way. Proximity to power shapes their
approach. They tell themselves a story about pragmatism to
justify political constraints, aware that pushing too hard could
mean losing funding or access. Their strategy is to make reform
feel politically safer to decision-makers, by not demanding too
much of them. This explains why political leaders so readily embrace 'trauma-informed' approaches in schools, prisons
and mental health services – language shifts while meaningful change remains elusive. The rhetoric provides the illusion
of progress while resources for real transformation are with-
held. Both radical and pragmatic approaches have their place,
their victories and failures. But in both, lived experience is
wielded as a political tool. People sharing their stories often
believe their testimony will drive change. More often than not,

however, the storytellers acting as the Trojan horse in many of these efforts will have little sway over what that change might actually look like.

*

Still, we must not abandon hope entirely, tempting a prospect as it is. Many reject traditional activism, joining the fight by other means – most often through social media. Yet, driven by unprocessed trauma, many throw themselves into battle, believing conflict is where healing happens, unaware of how their wounds fuel their actions. Trauma and politics frequently intersect, and when faced with fight, flight or fawn responses, many choose to fight without thought – regardless of their readiness for combat. The state of public discourse reveals how unprocessed trauma escalates conflicts, trapping communities in rigid identities and distorted victimhood narratives. As discussed in the previous chapter, trauma does not remain static. Unintegrated, it lingers, mutates and retransmits, embedding itself in identity, relationships and worldview. It shapes how we see ourselves, interpret systems and engage with the world. Left unaddressed, it ripples beyond individuals, influencing families, communities and even the nature of activism itself. Across societies, an unprecedented surge in activism and cultural conflict has emerged, with groups rallying around singular issues, convinced that eradicating one perceived threat will restore safety and order. While many of these causes address real injustices requiring deep consideration and careful mediation, the intensity with which political objectives are pursued often signals something deeper – activism as a trauma response. When we feel unsafe or unseen, we look for external explanations, tangible enemies and solvable problems, believing resolution will bring peace. But this belief is often misguided. Trauma does not subside once a political battle is won. Without healing, fear, insecurity and mistrust persist, projecting onto new targets. Like broken relationships doomed to repeat the same patterns, political movements too can become trapped in cycles of grievance, continuously shifting focus from one perceived threat to the next. While trauma is real, injustice exists and reform is necessary, activism that does not allow space for agency, healing and transformation remains stunted and inefficient. Today, I detect a mean-spiritedness in much of the campaigning I observe, particularly online, where various tribes have internalised and sanctified the notion that their opponents are not simply disagreeable or wrong, but are acting dishonestly with malevolent intent. Once this powerful notion is acquired and becomes operative, the traditional campaigning calculus is tossed aside. The aim of the game is no longer simply to advance on the strength of arguments and evidence, but also on how effectively you can smear and dehumanise the out-group. Whatever political issue currently preoccupies you, surely you recognise some aspect of this behaviour in your own conduct. In my view, this is a particularly post-modern trend. One that cuts across the political spectrum, which is present most palpably wherever social media and untreated trauma converge. Perhaps I reminisce with rose-tinted glasses, but the campaigners I grew up around were memorable not for how angrily or stridently they behaved, or how sneering or cruel they could be, but for how magnanimous, measured and strategic they were. I recall their open-mindedness and skill in braiding once disparate, often antagonistic sections of the community together. Their talent at bringing previously apathetic people along with them. Their cooler temperaments and their ability to make powerful persuasive arguments where criticism was seen as an opportunity to refine and clarify – not as an existential threat – are what stayed with me. Their ability to hold complexity, to acknowledge that they may not have all the answers, to engage with nuance – these are crucial for meaningful dialogue and progress. Their campaigning style was informed not only by the political issues at hand but by remaining attuned to the undecided observer looking on from a distance, and what they might make of it all based on how activists carried themselves. This way of doing things was modelled by leaders whose knowledge was not confined merely to politics – they knew something of themselves, too. That self-awareness granted them the gifts of empathy, affability and self-regulation, which rendered their politics all the more attractive a prospect. Today, we campaign on social media mainly with the faithful in mind. We hold ourselves in such a way as to be seen favourably by those who already agree with us, because this creates a sense of safety. We appear to the undecided or casual onlooker like we believe every issue is cut-and-dry, and that anyone who fails to see the issue at hand as we do is either a moral reprobate who needs to educate themselves or a bigot worthy of immediate cancellation. This is a far cry from the pre-digital age and, in my view, all movements are generally the worser for it. Of course, in the 1990s there was plenty of trauma, but we had no social media. Bonds were forged in the real world. Solidarity wasn't just a hashtag or platitude; we made those values visible by not only affirming one another but also by holding ourselves accountable when required. Yet when activism is driven by an activated nervous system, in a distinctly digital age, these principles are often

abandoned. Worse, rejecting them is seen by some as a necessary condition of political participation. Winning no longer
satisfies; opponents must be humiliated and defamed. Ironically,
it is often when we internalise powerful narratives of victimhood without parallel work on our wounds that we become
likelier to conduct ourselves in this unbecoming and politically
futile way. There is a crucial distinction between victim status
and a victim mentality, which we covered earlier. Activism
rooted in victim mentality offers both comfort and constraint
– it provides clarity of purpose and bonds of solidarity but can
also obscure toxic narratives and behaviours. Despite deep
ideological divides, when observed from a safe distance, war-
ring factions behave in strikingly similar ways when driven by
unprocessed trauma. In-group experiences are prioritised,
while out-group suffering is dismissed as fabricated or politically motivated. Out-groups are framed not just as ideological
opponents but as threats to in-group existence, rendering compromise impossible. If we took many recent campaigns at face
value, nearly every cultural conflict is a zero-sum game –
winner takes all, with no room for concession. This is not
political debate. It is large groups of activists, campaigners and
followers operating from deep woundedness. Their legitimate
anger is often directed or weaponised by political and media
actors and juiced by the cynical mechanics of tech platforms
for their own strategic purposes. Until this is recognised and
addressed, the cycle will continue – perpetuating conflict, reinforcing division and leaving the root causes of suffering largely
unexamined. Trauma wreaks havoc on our threat sensitivity
and risk perception, amplified by algorithms calibrated to keep
us trapped in a state of vigilance. From such unstable ground,
with so little account taken of the bleak digital context in

which activism is now primarily situated, how exactly can any movement for meaningful and lasting change be mounted? This is not a dismissal of activism or advocacy, nor an attempt to police tone or approach. This is simply a call to greater self-awareness of the role unprocessed trauma – aggravated by a technological revolution – may play in our efforts to bring about the justice we deserve. Whatever we are fighting for, we must ask: are we fighting for change because we have healed – or because we have not? Are we motivated by hope, or by unresolved pain? Do we wish to lift up our fellows, or tear down our abusers? The answers matter, they really do. If you move into position on the battlefield without first having considered these questions, you'll quickly become disorientated. How do we feel physically when we engage in our activism? Do we experience a sense that we are grounded, protected even – the body's way of reassuring us that we are on the right path? Or do we conduct our business while drowning in our emotional floods – heart rate elevated, laboured breathing, feeling like the walls of our lives are closing in? Do we disagree with our opponents while remaining aware they are people like us, often with as rich and layered an internal justification for their views as we do, or does dehumanising them make our job that bit easier? Some of you will inevitably view my stance as fence-sitting or both-sides-ism, but could any of you name a long-standing conflict that was resolved in any meaningful or lasting way which did not, in the end, involve compromise of some kind for both parties? I know all too well the invigorating effects of righteous anger and how the world feels when viewed through a thick crimson mist. I could write a book about all the people I've lost and gladly throw it at you – I don't make this argument from a place of privilege or remoteness.

I am well aware of the labour involved in not only fighting for change, but also in aspiring to maintain some semblance of perspective, discipline and intellectual honesty in the trenches. When everyone else appears to have abandoned those principles, what's the point in even trying? Often, it's the painful proposition of compromise with people we come to regard as sub-human that makes the bruising war of attrition feel like the easier path. We must realise that through a digital lens, much of the terrain we think we are observing is, in fact, an illusion. This is especially true if we carry untreated trauma. Social media makes us *feel* like we're winning – that's how algorithms induce us into playing their games – so what's the incentive to lay down our arms and talk? Whatever world is eventually forged from our place of woundedness, it will not be a happy one, and whatever successes are derived from our Pyrrhic digital victories, they will not last, I assure you. Trauma-driven activism, while powerful in the short term and satisfying in the white-hot light of conflict, eventually burns us out. It fractures movements, alienates potential allies, unites previously disparate factions against it, and leaves its participants emotionally exhausted and fearful – whatever the cause, wherever the fight. Healing-driven activism, by contrast, is rooted in resilience, connection and a clear-eyed understanding of both personal and collective power. It acknowledges the role of systems and structures while also holding space for the perspective of the other. It recognises that true change occurs not simply when beliefs are being asserted with conviction, but when that parallel work on ourselves, in the context of safe and nurturing relationships, is also being done.

*

In today's climate, it may seem obtuse to ponder whether changing the world is even possible outside the political sphere. Many would argue no, but I happen to disagree. When I engage with people in recovery, my priority is to offer hope and support for sobriety unconditionally – not to politicise pain or co-opt their experiences. Had I been screened for my political beliefs when I stumbled through the doors in 2013, I suspect I'd have been thrown back onto the street. Countless men I've supported in the early days of their recovery were riddled with grievances and resentments entirely antithetical to my personal politics. I welcomed them into my home nonetheless, and we did the work. That's the work that keeps me well and that's the work that makes me more effective whenever I do choose to enter the political fray. It's when I stop doing that work that I come undone under the weight of my own nonsense and my activism becomes toxic. I work a clear boundary between my recovery and my activism; you'll rarely hear me speak on political issues in my recovery groups as doing so would threaten their integrity. Healing in this context arises from mutual aid, respect and relationships based on trust – not lofty political goals. In those rooms, we leave our politics at the door. It's precisely because we keep politics and ideology out of those sacred spaces that healing is even possible. Addicts are one of the most persecuted and marginalised minorities in society. We experience discrimination in health and criminal justice. We are mistreated by professionals who fundamentally misunderstand our mental and physical condition, and all of this is exacerbated by great prejudice and ignorance of what we suffer from. Each of those systemic injustices costs thousands of lives every year and has firm political roots. Yet, however earnestly we desire change, politics by its nature activates both the ego and the nervous system – not ideal in the early days of recovery. We set aside our anger and try to accept the world as it is for a time. We recognise that our beliefs and identities, and the stories we've been telling ourselves about who we were and why, were often mirages produced by our alcoholic projections, themselves fabrications of untreated trauma. We come to learn there is an order of business and that changing the world may have to come later, if at all. While systemic reform remains essential, and throughout our lives we may be called upon to enter the political sphere to fight for justice no matter our fitness for battle, the idealism underpinning much of trauma-adjacent campaigning often inflames the bodily condition of trauma, running contrary to the principles of healing, at least as I understand them. I see influencers and activists online, coaching clients on how to ascertain the political beliefs of their therapists, ensuring they don't encounter anyone who might see the world differently, as it may be retraumatising. I see lived experiences politicised before the individuals called upon to share their lives are even aware of their own views or opinions on political issues. And I see campaigning organisations, which purport to educate us about basic principles like coercion and informed consent, extracting narratives from survivors (who often lack the necessary experience to truly exercise autonomy) thrust into the spotlight before it's clear whether they are in a fit state to cope with the fallout. As discussed, healing from trauma involves developing a willingness to toss aside certain incomplete narratives which no longer serve us. Yet, in the realm of campaigning, we are often asked to embrace falsehoods in service of the ambitions of third parties, be they radical or pragmatic. It may be far more useful to regard any political activity in which we may engage as supplementary and entirely separate from our recovery work – not a precondition of it.

While it is simply undeniable that trauma has deep systemic roots and requires political action to address (and I will continue to engage on that front as I have always done) I've been moving in this radioactive sphere for decades, and I can assure you, it's not the place to go if you want to get or stay well. Healed people do not play games with human lives. They don't extend or withdraw basic compassion or empathy on the basis of another's beliefs or background. And they don't make political capital out of the trauma and grief of the most vulnerable or play one group's pain off against another's for reasons of strategy – these are games that wounded, toxic people play. The hardest fact to face is that when we're behaving like this, we often have no idea we're doing it; this behaviour becomes second nature, reinforced by assigned identities and incomplete narratives, turbo-charged by algorithms designed to confine us to curated parallel realities. Too many 'leaders', whether in politics, civic society or at a community level, have simply normalised deeply dishonest conduct that would be deemed unacceptable in most households. They don't just lie to you, they lie to themselves. Far from lacking understanding or political consciousness, the further into recovery some of us get, the less time we might want to spend around people like that. Those in recovery who consciously decline the invitation to participate politically on such terms, far from not caring, often understand the politics of pain perfectly well, they simply choose to stay out of it at the public level – like alcoholics stay out of cocktail bars. The way through, as I see it, is for people living with the scars of trauma to recognise that change can come in many forms and need not be led by the politically minded radical or pragmatist, nor the opportunistic politician. It needn't be political at all. While healing cannot occur outside the context of relationships, a nurturing relational network can be entirely free of political motive – don't let anyone tell you any different. Your energy may be better spent running as far from political campaigning as possible, and strength to anyone who chooses to do so. Despite our hopes for a better, more compassionate world, the reality is sobering. That's why those of us living with trauma who wish to see a fairer society need to wise up. The evidence suggests that our ideal society is far from materialising, and operating from woundedness while attempting to participate politically may be our small part in that agonising stalemate. Every minute we spend unconsciously acting from our wounds, we risk playing into the hands of others who would co-opt or reframe our anger for their own purposes.

In many ways, political participation in the digital age mirrors the dynamics of a dysfunctional family, and we too often play the roles assigned to us. In such families, toxic members dominate, dictating how everyone else must think and behave. They justify their tyranny through deception and coercion, ensuring that any challenge to their authority is met with punishment or shame. Over time, the rest of the family internalises the blame – or goes angrily in search of a scapegoat. In our search for safety, we become more open to the suggestions of these manipulators and triangulators who claim to represent our interests, but whose true aim is often to influence and exploit us for their own ends. They position themselves as saviours or allies while sowing division, feeding our insecurities and directing our pain in ways that serve their agendas. In this conduct, they present a deeply distorted model of what it means to fight for change. Isn't this the truth of it, when you take a clear-headed view? It only takes a handful of toxic actors to curdle an entire debate, yet the more thoughtful among us carry this burden, questioning ourselves and our commitment to the cause, tempted to tweak our identities to fulfil the expectations of the faithful, instead of recognising the dysfunction and the part we unwittingly play as it unfolds. I have played my part in this charade, gratifying my fragile ego at the expense of nuance, complexity and even truth – as have many of you. This realisation can be unsettling, even painful, but we are all victims and perpetrators of something in the end – if that offends, then whatever you have suffered, you have yet to recover. Let this be a moment of awakening. To see the dynamic for what it truly is – parallel discourses structured like toxic family systems – is to begin disentangling ourselves from the incomplete narratives and assigned identities that constrain our growth and usefulness. This is not a call to despair but to clarity. But this awakening must also encompass an understanding that irrespective of how unjust our world may be, and how much it needs to change, much of the work of healing the wounds it inflicts remains ours alone to perform. For some of us, me included, trauma-driven activism feels like a safer, more familiar path, easier to tread than the unlit and overgrown road of self-examination. Some would sooner tear their precious movements apart from the inside, acting recklessly from their woundedness, than just admit they've lost the plot and go to fucking therapy.

*

This shouldn't be taken as a green light to descend into the individualist rabbit hole, endlessly lapping the Trauma Industrial

Complex in a tedious loop of awareness building. In the end, trauma is both a personal and collective endeavour and we can't do it alone. Recovery requires a delicate balance of agency and community, of self-reliance and support. It requires that we hold two truths at once: we are shaped by our circumstances, but not entirely defined by them. That we are victims of systems, but also the authors of our own lives. I am always being reminded by my comrades that only a political revolution, and wholesale overhaul of the system, will ever be sufficient in raising the quality of life of those who languish in precarity and poverty and exclusion. But even in the event of our hypothetical utopia becoming a reality, where a service rises to greet every unmet need, and every resource is made available to tend to every ailment, all the experts and money and time in the world will not prevent an alcoholic from drinking until they decide to stop. This is evidence of agency at the level of the individual – agency which, when channelled correctly, may have as transformative an effect on a person's circumstances as entourages of wrap-around professional support. A more radical communitarian approach may be required in light of the cold, hard truth that for some, help beyond their most basic needs simply is not coming. In my community, where vulnerable people require urgent assistance, it seems to me that pursuing lofty, often abstract political goals, or roleplaying as revolutionaries online are arguably the greater indulgences. We live in a time of immense turmoil, uncertainty and hardship, which requires more of us than merely surfing the self-help aisles, crafting designer identities or making up the numbers on someone else's political team sheet. We must rise up as individuals and take responsibility for our healing, and then, if we desire, leverage our newfound power to bring about change – not in a wider abstract world that barely registers our existence, but on our own bloody doorsteps. Love can be a profoundly political act. But what if love is about more than just putting your arm around someone and hearing their story? What if love, in a deeper sense, is also about looking them in the eye and telling some home truths? Truths upon which their life may someday depend. As I see it, one truth some of us may have to accept, despite our idealistic reservations, is that we may not be able to change the world. Or rather, the change we're sold is unlikely to arrive in the grand, transformative manner many theorists, academics and campaigners envision. I can feel my assigned political identity lashing back from within as I put into words what would have been previously unsayable to me. But I am not making a case here that change isn't possible, or that a better quality of life lies permanently beyond reach. Nor do I encourage using the immensity of the current moment as an excuse to become apathetic or nihilistic. I simply mean that we must whittle our hopes down to a more manageable size while tending to our wounds, relieving ourselves of grandiose designs and paying closer attention to the order of business. While certainly not as glamorous a notion as reordering the world around us in some idealised, self-aggrandising revolution, it's in our own broken communities that our experiences are often most urgently needed. It's on this front that we may come to find our true calling and perform our service to the world – whether it changes or not. Did getting sober solve every problem in my life? No. Did it offset the vast socioeconomic inequalities I faced at different points? No, of course not. But by prioritising the most urgent problem – addiction arising from untreated trauma – and becoming resolved to confront it,

I was, after a time, better placed to bring myself more fully to the many other challenges beyond my immediate control. And as my personal load lightened, my attention gradually turned to how I might help others. In all my years of campaigning on class inequality, trauma and addiction, very little has changed in the world of politics. It's in the rooms of recovery – where no facilitators are present, where politicians dare not venture, but where stories are shared freely in a safe, discreet and genuinely loving environment – that I bear witness to true radical transformation. And all we do in those rooms is put out some chairs, boil a couple of kettles, and humbly hold space for others like it was once held for us. In the end comrades, for some of us, that's what changing the world looks like.

'Caring for myself is not self-indulgence, it is self-preservation, and that is an act of political warfare.'

Audre Lorde, *A Burst of Light* (1988)

The Freedom of an Unwritten Future

*How do we embrace authenticity,
let go of attachment and move forward
into a new chapter of our story?*

It's a chilly December morning in Glasgow's West End, and some of my toughest experiences in life are about to be put to use. A pioneering initiative to educate trainee GPs on the subtler aspects of addiction – Humanising Healthcare – has brought together future doctors and people like me, in recovery. I am here to open the event with my disclosure – what we in the recovery community call a 'share'. I begin with my upbringing – poverty, trauma, class – authenticating myself in the eyes of the other participants, who have walked similar paths. Then I recount the progression of my alcoholism. It started on the day of my mum's funeral, when a relative she occasionally used drugs with handed me a frosted bottle of beer. I remember standing in his living room, wearing a long dark coat, black suit, and tie, feeling, for the first time, like a man. I'd tried alcohol before and it had never grabbed me, but this drink was different. It made me feel something – something better than the numbness. So, I had another one. And then another. Within a year, the man who handed me that beer was dead. Within five, I was a full-blown alcoholic. In early recovery, my shares focused on the causes of my addiction or the chaos that followed. They focused on the problem. These stories became the foundation of my sober self. But as with all man-made structures, time brings wear and tear. Eventually, repairs, renovations or even demolition become necessary. As I grew, my shares evolved. I began to examine my own role in my suffering, acknowledging the places where I was not just a victim but also a perpetrator of harm as well as a collaborator in my own difficulty. Each time I shared, whether in a meeting or at an event like this one, I walked away with a greater sense of peace – the nervous system's reward for choosing truth over comfort. My shares began focusing on the solution. That peace is in stark contrast to the turmoil that comes when I choose comfort, or the promise of fleeting gratification. When I choose public visibility for its own sake, and the expectation that I must leverage my platform in a specific performative way. The voice inside me insists that I must be seen to hold people accountable, engage in verbal combat, and never be caught empathising, or even fraternising, with the out-group. No one has explicitly told me I must play this role, but the voice berates me whenever I dare to consider saving my time and energy by ducking out of a particular social media storm or stepping away from battle altogether. Worse, it whispers that all my years of fighting have yielded little beyond a bolstered ego and conditional acceptance from a tribe that loves me only as long as I play my assigned role – the working-class guy who made it but never forgot his old arse. This voice speaks whenever I take the stage. It speaks when I open a blank document to write. I once mistook it for my fundamental essence, my calling. But so much of what this voice demands leaves me anxious, fearful and exhausted. As I stand before these trainee doctors, that voice sneers, reminding me that what I'm doing will never make a difference, that this just isn't radical enough. But each time I ignore that voice, as I am choosing to on this morning, it gets a little quieter. And each time I muster the courage to tune it out, a surge of gratitude washes over me – my nervous system reassuring me I'm on the right path. That same peace washes over me when I sit with another addict and let them do the talking. It overwhelms me when I set aside excuses that I'm too tired and too busy and just give my children my full attention. It captures my soul when I hold my wife and love her in a way her nervous system can understand – not the way that comes easiest to mine. And I feel that peace when I set my thoughts down on the page, not because I feel obligated to, but because I have something to express that rings true to me. Long after trauma granted me the temporary superpowers of self-preservation, it condemned me to the misery of self-soothing selfishness. In that state, my inner voice fixates on my own needs, my entitlements and how others should accommodate them or be punished – unrealistic expectations of the anxious little boy inside me. That boy, in his endless search for safety, pursues attachments that offer fleeting relief: validation, wealth, material objects and carnal pleasure. But the momentary sense of security fades like a footprint in the driving snow, and the pursuit begins again, looping endlessly towards self-destruction. For me, that self-destruction lies in living in an incomplete story. Today, at the university, using my experience in a safe and practical way – to help health professionals better understand the plight of addicts – I may not be leading a revolution, but I feel something approaching authenticity. My nervous system brought me here and it's begging me to love the anxious little boy inside me, by telling him a new story.

The notion of the authentic self, so prevalent in the trauma field and self-help, has always confused me. In storytelling terms, it has served as the ultimate MacGuffin driving my narrative forward – in reality, just another spooling reel. Much like the hunt for a magic stone or a long-lost key to a secret door, the desire to obtain this mystery item propels a protagonist through the beats of their tale, but unlike fantasy novels or sci-fi epics, the search for my authentic self never seems to end. People confuse my openness, anger and honesty for authenticity. My willingness to speak out is, for some, a sign of character. Yet every time I dive headlong into a fight, on whichever front it appears that day, I am often resisting what feels truly authentic in favour of pursuing attachments. There is my attachment to the idea that I may finally experience acceptance and redemption by fulfilling a role I assume others expect of me. There is my attachment to the dream of financial security, which thrusts me into the public eye whenever my work must be publicised. And there is the stubborn attachment to the belief that if I heed my nervous system and choose a quieter, less opinionated, less visible life, others might judge me harshly. The contradiction at the core of my life is that the persona I have co-created with culture represents only a slice of who I am and what I care about. The truth is, in my pursuit of how I am perceived, my security instincts and my fear of criticism, I abandon the prospect of inhabiting my authentic self – if such a self even exists. Authenticity, for many of us, is just another performance. The only time I experience true relief from anxiety, dread and fear is when I focus on others rather than myself. And every time I abandon my true need and desire – to lead a quieter, more community-focused life while raising my children and remaining sober – in favour of pursuing attachments – I am driven not always by selflessness, but by a subtle form of self-seeking masquerading as altruism. One of the biggest barriers to healing arises when we realise that we must let go of parts of ourselves that no longer feel true. We begin to sense this misalignment between our authentic selves and our attachments through our bodies – if we truly pay attention. This pull is deeper than being triggered; it is a knowing in our bones that some aspect of our life is fundamentally at odds with who we truly are. Having made a name for myself as a vocal campaigner on social issues, my audience wasn't just drawn to my arguments but to the intensity with which I delivered them. My righteousness validated their own sense of injustice and desire for accountability. What many didn't realise was that I was either in the grip of addiction or trying hard to overcome it, and my own internal conflict fuelled many of my strident calls to action as much as any external injustice. A few years into sobriety, I began to calm down. Rather than using anger as creative or political fuel, I started viewing it sceptically, as a red flag rather than a green light. I underestimated how this shift would affect my social standing. The more measured I became, the less useful I appeared to the cause. Many in the hip-hop community that raised me came to regard me as washed-up. Old school friends publicly contradicted my story, as if I'd fabricated my upbringing. And by today's standards, in many former comrades' eyes, I am a terrible activist. I rarely raise my voice, and I take too long to consider most issues before signalling a view – if I signal one at all. Heeding the advice a seasoned campaigner offered me many years ago, I have learned that making my values visible needn't always become an act of public exhibition. Choosing not to air a half-baked view on the trending topic of the day may, in fact, be a simple act of self-care – not everyone has earned the honour of knowing everything I think, feel or believe. I must honour my truth that in today's climate, it feels neither safe nor practical to bare one's political soul publicly on a daily basis. Furthermore, having faced my own moral complexities, I am now less inclined to paint the words and deeds of others in stark black and white. For some, of course, this signals ideological impurity. Admittedly, I remain too vain to fully disregard the opinions of others. I swing between extremes – fantasising about unmasking my true self and accepting the fallout, then fearing that perhaps I have lost my way and that their criticisms may offer a way back to who I really am. I am trapped within the frame of my own self-portrait, an image formed by my unconscious drive to meet my needs through external validation. Like many who suffer serious trauma early in life, my sense of self is intermittent and unstable. To paraphrase Eminem, I am whatever they say I am: a radical, a sell-out, the authentic voice of a generation, a grifter, an Orwell for today's poor, a useful idiot and a working-class hero only in it for himself. These labels replay in my mind many mornings as I awaken and try to enter the world on terms that feel truthful. But embracing the person I am today carries consequences. It's not that my values or politics have changed, but my temperament and my once-grandiose belief that I could change the world in some profound, revolutionary manner have softened. Not because I don't want it to change, but because I now recognise a world created in accordance with my values and beliefs would feel like a hellscape to many others – including many of the people I think I'm advocating on behalf of. This realisation creates doubt, forcing me to pause. At some point, I convinced myself that my life had to be offered up to a cause. That for things to have turned out well for me, I must suffer. That in order to justify my success, I must struggle. And so, I have suffered – trapped in the pursuit of roles I think others need me to play, even when those roles contradict one another, or make me physically and mentally ill. I want so badly to be loved that I am willing to abandon myself in pursuit of safety and acceptance, even when it is fleeting or illusory. I'll self-censor. I'll feign anger where there is none. I'll say nothing when I'm truly raging. I'll pretend to be interested when I'm not. Sure, life still gets under my skin, and sometimes I still dive headfirst into battle before I've realised what's happened – that's still part of who I am. But the idea that my life must be offered as a sacrifice to a larger cause is pure ego, triggered by a fear of judgement. The truth is, I feel most useful at home with my family, or in my local community, performing service humbly and anonymously. Being celebrated now and then is nice, as is the thrill of debate, but I need people around me who will hold me accountable and draw me back out of my delusions whenever I grow tired of doing the work. Many trauma survivors unconsciously return to the scene of the crime, hoping to rewrite the script of their wounds. We go back to relationships that broke us. We return to dangerous environments that put us at risk. We show up in our lives in ways that guarantee our circumstances will not improve. I've been trying to break this pattern – sometimes succeeding, sometimes failing – but even in failure, I've learned enough to know I'm once again on the right path. After a year spent writing about trauma while recovering from a deep traumatic wound, it's time to confront the attachments I've been afraid to let go of. This work is now well underway. A few weeks ago, I retired Loki – the name I gave to that little boy – with 300 people in attendance for my final album under that moniker. It was a symbolic break from a cherished but outdated part of my life. I took the moniker of Loki as a wayward teenager. It gave me a chance to create a new identity, a bit like Jim Carrey's character Stanley Ipkiss in *The Mask*. As time went on though, and I got more known in the scene, and deeper into alcohol and drugs, I now believe I became stunted emotionally. Loki represents a part of me that's frozen in time. An entity I summon when unavoidable conflict arises, and I require a different set of tools and skills to navigate it. Loki is less considered and cautious, quicker to rise to the bait, and views every battle as a zero-sum equation. Admittedly, he's probably more interesting and exciting than I am, but he also cares far too much about what other people think – Loki is my trauma personified. He wasn't the only attachment I decided to let go. It's time to make an exit from the hot-take culture currently passing as informed commentary. There is quite simply no way I can dedicate the energy necessary to stay across every trending topic. Pretending otherwise does a disservice to the issues at hand as well as my platform. Algorithms demand that we all become generalists, confident to speak freely on any subject irrespective of our experience or qualifications to do so. While everyone is entitled to their opinion, a public profile must come with certain responsibilities. I refuse to contribute any more hot air than is absolutely necessary to debates dominated by shock-jocks evidently more fascinated with the sound of their own voices than they are in engaging nuance or creating understanding. Every time I speak, I suck oxygen out of the room that may be better utilised by someone more knowledgeable than I on a given topic. I gave up my podcast, a long-held ambition of mine, recognising that its long-term viability required a level of commitment I could no longer justify, and that my motivation for starting it was to set up shop permanently in a highly visible political space – my nervous system's idea of hell. I'll remain as close to my home turf as is practical from now on, rather than pitching myself as an omnipresent talking head on current affairs. I will continue to decline most media requests, large and small, and try to resist being drawn into white-hot debates about trending culture-war topics unless I genuinely feel I have something of value to contribute. It seems many have built their entire identity around these issues now, and while I would never dismiss the seriousness of the various debates comprising these conflicts, I've decided those digital trenches are fronts I no longer wish to engage on so frequently. But healing isn't just about cutting things out of my life; it's also about what I am willing to build in. For me, that means making more time for others. Over the past year, I've increased my recovery service, not just for my own wellbeing but as an expression of gratitude for what was freely given to me – my sanity. Being well isn't just about seeking help when you're struggling; it's about being there when others need you. I now dedicate seven hours a week to various recovery commitments – more than I ever have. As well as acting as treasurer for one group and secretary for another, I also work a shift on a helpline once a week, taking calls from across the UK and directing addicts in need to local responders and services. A lot of the time, truth be told, I cannot be arsed, but I rarely skive on these obligations and always feel better afterwards. I also became a patron of

EMDR UK – an association of mental health professionals pushing for wider access to trauma therapy. The role was formally offered to me by Marlyn's husband, Mike, also a psychologist, and after giving it some consideration I jumped at the opportunity. Not just to give something back but for the chance to see a little more of my old friend – Marlyn tags along to conferences, apparently. In the long term, I hope to expand my work on trauma and I have even designed and piloted a new programme of workshops that aim to help storytellers construct narratives with greater intention, and facilitators to elicit those stories more ethically, with duty of care at the heart of the process. I've also accepted that no amount of money in my bank account brings me peace if that anxious little boy is running the show. With that in mind, I choose to remain home most of the time now, rather than aggressively pursuing work based on the fallacy that a great credit rating and happy children are one and the same. I'll take life as it comes, trusting that what's meant for me will materialise and what isn't will pass me by. And while I could never completely step away from my campaigning, my days of waging ideological war daily online, in the vain hope of earning a pat on the back from people who clearly couldn't give two fucks about me and contribute nothing but negativity to my life, are over. This next phase of my journey is about recognising how my attachments – to my incomplete story, to my assigned identity, to my skewed sense of victimhood – have prevented me from living a more authentic and fulfilling life. If I must carry a message, let it be one of healing, not one of ceaseless conflict. And if that doesn't work out, I can always construct another identity, book in for that long overdue ADD assessment and finally find out if Gabor was onto something.

What feels most real to me, beyond my family and friends, is the time I spend with others in recovery. Here, I sense true authenticity. We don't sit around discussing our problems; we are just as focused on solutions: we organise, plan and follow through on our commitments. There is no time for small talk because we know that to stay well, we must go beyond our public personas. And without addiction and trauma, many of us would never have crossed paths. Yet, despite our diversity, we rarely argue. Despite our hardships, we rarely wallow in our misfortunes. You might assume that what we do is insignificant in a world falling apart, but you'd be wrong. Our movement spans decades, it covers the globe, and the only reason it persists is because we actively resist the temptation to become embroiled in politics or contentious matters of public debate. We carry our message beyond the rooms of recovery, into institutions – hospitals, mental health facilities, rehabs and prisons. We are entirely self-funded, handling sums of money large and small, and every decision at group, local, national and international level is taken democratically. If activism is about transformation, where is that more embodied than this, or in the recovery of those who once used drugs intravenously every day of their lives? Where is true impact more evident than organising on a daily basis, to ensure help is always available, day or night, and walking alongside broken people until they become well? For years, I confused articulating a social problem with addressing it, just as I confused describing emotions for feeling them. The satisfaction of being praised for my words often took priority over the urgent needs right in front of me. Enough is enough. These are hard truths I can no longer afford to overlook or minimise. I am tired of arguing on the internet. I am tired of ignoring one tribe's faults while shouting about another's from the digital rooftops. I am exhausted by the endless stream of bad news, the curated realities that reinforce my worst fears, and I am sick of the person I have to become to play this game 'successfully'. The angrier and more spiteful I am, the more praise I often receive. And where has that left me in the past? Strung out in rehab. Rolling around in the foetal position in police cells. Battling mental illness. Facing family dysfunction and marital mayhem. Living with constant financial insecurity and fear of public condemnation. Afraid to even feel and as good as fucking useless. Some may say if I continue to heal, I may come to better withstand these pressures. But I now understand that these pressures arise from a life lived on entirely the wrong terms – a life lived inauthentically is what creates these pressures. I could deny what I know in my heart, and seek refuge in the Trauma Industrial Complex, whether online or in safe civic spaces. I could choose to remain confined by the frame of my self-portrait. The affirmation, sympathy and shock on endless supply, the promise of justice, dangled like a carrot as I dive time and time again into someone else's fight. The safety of my assigned identity as the working-class hero who never forgot his old arse, providing comfort, familiarity and acclaim. Or I could once again push forward, into the next chapter of my story. These wounds have continually driven me into situations I knew were unhealthy, hoping the deepest, wisest part of me, gnawing at the pit of my stomach, was mistaken. They allowed me to develop a masterful performance of self-awareness and insight that not even the experts could see was a forgery. Whatever consequences flow from stepping away from the aspects of this life that no longer feel real to me, and into a great unknown, I now embrace fearlessly: This is not an identity, it is a prison. This is not truth, it's a lie. This is not authenticity, this is an incomplete story. A story I must now rewrite.

'To live past the end of your
myth is a perilous thing.'

Anne Carson, *The Glass Essay* (1994)

Holding Ourselves Accountable

Conclusions and recommendations

Years ago, when the relapse that brought me once again to my knees was well in motion, I gave a keynote speech at Scotland's first ACEs conference and participated in a conversation with Gabor Maté – a few months after we met in London. In my speech, I didn't focus on trauma itself but on the need to treat the message – that trauma is real and untreated creates untold moral and financial costs – as a political project. I argued that we needed to engage the public in savvier ways and compel decision-makers to act. But I also offered the audience my observations of the trajectory of our rapidly growing global conversation, and my sense that we had to refine our approach. Many found my message compelling; others seemed confused. Lived experience people like me are often expected to stick to their narratives and leave the big ideas of strategy to the facilitators. Reflecting on that time, I felt hurt that some in the crowd seemed more interested in my personal story than in my political message. I foresaw a day many of them never thought would come. When backlash would force us out of our comfort zones to account for our assertions about trauma. I felt it imperative that we get ahead of that storm, by openly acknowledging that many of our 'facts' are theories, and many of our 'frameworks' are stories: Maslow's hierarchy of needs, the five stages of grief, the 12-step programme – all stories with great utility but stories, nonetheless. Perhaps I was trying to pre-empt a fight I felt was inevitable, given the culture wars were by then well under way. Still, I felt resistance to my straying from the familiar boundaries of personal storytelling as well as anxiety at my attempt to politicise what remains a largely apolitical discourse. I left that day and kept my distance. And in typical form, I began contemplating how I might one day return with a new story. One that would compel any doubters that morning, to take me and my message a little more seriously. A story that might persuade those people who always insist 'we must listen to survivors', to put those words humbly into practice. It remains to be seen whether I will succeed in that aim, or whether this book will gather dust on the misery-memoir shelves in their homes or offices. I suspect some will see the value in my lived experience of being lived experienced, and the sincerity with which it's presented. Others may come to regard this book as an attempt to upcycle my adversities once more, this time as a three-act hero's journey, complete with that happy, inspiring finale no tall tale can seemingly do without. The sceptics among you, meanwhile, may detect a sardonic meta-commentary on the trappings of trauma voyeurism, laying bare the rudiments of personal storytelling in a market driven by ambulance-chasing consumers. Perhaps it's none of those things, perhaps it's a braiding of all three. Whatever your thoughts or feelings, the story you go on to tell yourselves about this book is no concern of mine – I know why I wrote it and that's enough for me. In light of everything we have explored, I believe we must accept that trauma-related storytelling is here to stay, it will vary in quality and utility, and that the best defence against the risks posed by the Trauma Industrial Complex is a more robust culture of safeguarding – a culture led by lived experienced people themselves. A culture that pays more than lip service to the wellbeing of people with trauma. Operating from wounds, we leave ourselves open to further adversity. Confusing the telling of a story with actually healing from trauma is a risk many of us cannot afford to take. The best defence against opening this can of worms is to move past the awareness-building phase and into truly processing and releasing our pain. For some of us, that means actually healing – something few of us want to talk about because telling the story is much easier. From observing people like James and countless others I'm privileged to know who seem genuinely well, I've noticed one trait they all possess: self-awareness. They understand their baser natures and tend less to see their lives in black and white. They reject the world of heroes and villains. They are quick to admit wrongdoing and make amends. They understand how the human drive for prestige and security can cloud our judgement. They recognise when their outward kindness masks an ulterior motive. And when struggling with a problem, they aspire to identify and then pull it out by the root, rather than opting for an easier fix – like deceiving themselves. Ultimately, they work hard to remain aware of their own thinking and in doing so become less likely to behave impulsively or become dominated by emotions. It may seem unfair that, having been wounded by trauma, we must also carry the burden of confronting how those unhealed scars show up in our day-to-day lives. With trauma, there is rightly much focus on what we survived and of whom or what we were victims. If you're trying to get well online, that's exactly where a lot of content creators want to keep you – unhealed, lacking insight and highly malleable to their suggestions. True therapy not only involves affirming and validating our thinking, feelings and experiences, but also – just as often – asks us to let go of our old ideas about who we are, what happened and what it all means for our future. I'm only halfway there, but with the year I've had, and the wounds it has reopened, I find myself suddenly struck by a desire to push forward – writing this book for you has helped more than words could adequately express. For that, I thank you.

All that said, it seems appropriate to offer some suggestions to those of you who may feel moved into action as a result of reading. What follows is a brief shopping list geared towards anyone involved in the lived experience movement, whether facilitators or storytellers, as well as consumers of our stories. This is a marketplace like any other, and some tighter regulation is now required: Safeguarding the integrity of storytelling requires a commitment to ethical standards that protect both the storyteller and the stories themselves. Organisations and facilitators must develop clear ethical frameworks that guide how personal experiences are elicited, shared and used. Informed consent must be at the heart of these processes, ensuring participants fully understand where and how their narratives will be used, with the right to withdraw their testimony at any time. There must be a duty of care, including pre- and post-disclosure support from trauma-informed professionals, to mitigate the risks of re-traumatisation. Furthermore, those who tell their stories should not be expected to do so for free under the guise of raising awareness or giving back. Institutions that invite individuals to share their personal histories in research, training, advocacy or media contexts must ensure fair compensation that reflects both the emotional labour involved and the risks of putting themselves out there. Safe working conditions should be guaranteed, with provisions for breaks, the right to refuse certain topics, and access to mental health support. Training should also be made available to both storytellers and facilitators. Those who share their experiences should receive guidance on how to do so safely, with a focus on setting boundaries, emotional regulation and self-care. Facilitators, in turn, should be trauma-informed (preferably with first-hand experience of trauma and sharing publicly), trained in ethical listening, and committed to centring the storyteller's wellbeing rather than prioritising audience impact. Participants should retain agency over their narratives, including the right to remain anonymous if desired, and should not be confined to rigid, predefined arcs that reinforce a singular notion of what a survivor's story should be. Aftercare must be integrated into all storytelling spaces. Emotional support, structured debriefing and access to counselling should be standard practice, with organisations ensuring that follow-ups take place after public disclosures. A cultural shift is needed, where stepping away from public storytelling is not seen as a betrayal, but rather as a sign of healthy boundary-setting. The pressure to relive past trauma repeatedly, for the benefit of institutions, media or audiences, should be actively challenged and resisted. A key aspect of reforming the space around lived experience is shifting the power dynamic so that those who share their stories are not merely witnesses or case studies, but leaders and decision-makers within the organisations that claim to represent them. The current narrative economy is often dominated by facilitators and institutions, with those who hold direct experience relegated to the margins. Funding should be directed towards lived experience-led projects, businesses and advocacy efforts, rather than prioritising external researchers and consultants who study these communities without lived investment in them. Personal experience should be leveraged in ways that empower the storyteller, rather than reinforcing cycles of public victimhood for external consumption. Just as ethical consumerism has shaped industries through practices such as Fairtrade and sustainably sourced products, audiences engaging with highly personal narratives must develop similar discernment. Consumers should ask critical questions: Who benefits from this story being told? Has the storyteller been fairly compensated? Does this narrative centre the dignity and autonomy of the person sharing it, or is it being exploited to serve an external agenda? In the same way that people now challenge exploitative labour conditions in supply chains, we must interrogate the industries that commodify trauma – whether in media, publishing or advocacy spaces – without accountability. Ethical storytelling should be transparent about its process, much like ethically produced goods include information about their sourcing. Did the storyteller retain control over their own narrative? Were they supported throughout the process? Audiences must also reflect on their own emotional responses: are they engaging with these stories because they want to feel informed and empowered, or are they consuming them through a lens of voyeurism, pity or shock? I think we've all been guilty of the excesses outlined in this book, but while we must hold ourselves accountable for that, it's equally important not to dwell – not when we can learn from it and move forward. And finally, for those who share their themselves so freely, it is crucial you take active ownership of your narratives. Storytellers must ensure that their accounts remain truthful to the best of their knowledge, relevant to the context at hand, and responsible with regards to their own safety as well as the wellbeing of others connected to the disclosure. Before sharing, it is worth asking: Is this version of my story still accurate, or am I repeating a past version because it is expected of me? Does telling this story serve a meaningful purpose, or am I being pressured into it purely for impact? What are the potential consequences for me and others connected to this story? Storytelling must be a choice, not an obligation. Many feel a duty to share, but self-preservation must take precedence over audience reception. Additionally, people with lived experience must recognise their role within a broader system and advocate for themselves as workers within this space, much like members of a union. They must push for fair compensation, set boundaries around how their stories are framed, and resist exploitative practices such as unpaid engagements, unsafe environments and manipulative storytelling techniques or demands. Understanding the transactional nature of campaigning is essential, as many organisations that claim to centre lived experience often prioritise impact metrics, media engagement or funding streams over the wellbeing of those sharing their stories. Becoming comfortable with declining opportunities that compromise safety, ethics or autonomy is a necessary act of self-respect – and solidarity. By collectively organising, people with first-hand experience, whether storytellers or otherwise, can demand better conditions, ensure their voices are valued rather than extracted, and reshape the industry towards genuine empowerment rather than performative advocacy. Lived experience must not be treated as an endlessly extractable resource, but rather as a perspective that holds power, agency and the right to ethical treatment. These vital perspectives must be valued and protected – not just by institutions, but by those who share their stories as well. These recommendations aim to end exploitative storytelling, reduce harm and protect those sharing their experiences, in whatever context they are deployed or divulged. By encouraging conscious storytelling consumption, we guard against the pornification of trauma and ensure storytellers are protected and respected. Ultimately, however, the best guardrail against bad practice is for storytellers ourselves to take greater control of our narratives, keep ourselves safe, and thus begin modelling a better way of holding ourselves in this brave new world of trauma. I hope this book will make some small contribution to this effort.

Part of my aim in writing this book is to model a way of not simply telling your story, but of contemplating its true nature, and some of the implications of putting it out there. So much trauma-related content, and catharsis culture purporting to promote awareness, feeds not on the genuine desire to heal, but on the very symptoms of trauma outlined throughout this volume: high-impulsivity, poor boundaries and a desperate need to make sense of pain through stories. By embarking on a genuine process of healing, we can begin to accept the broken parts of ourselves we once hid out of shame or fear of judgement. We start learning what we can and cannot tolerate and rearrange our lives as a result. When we begin to see the benefits of this new way of being, we develop faith in our intuition when faced with important decisions. I've had a taste of this way of living, which is partly why my backslide into trauma has been so frustrating. These skills, when nurtured, act as guardrails against the magical thinking that compels so many of us to not only run with incomplete stories about our lives but also to hastily broadcast them publicly on misplaced assumptions that it will help. I won't claim to be a paragon of healing, but I have faced some hard truths about myself over the years. I couldn't have got sober otherwise. I am now resolved to face some more. For now, it's worth emphasising that while wisdom from friends like James and Marlyn, or kind and compassionate professionals like Suzanne, Miriam, David and Marie are invaluable, healing requires more than contemplation or storytelling – it requires action. Without it, we risk staying stuck in the awareness-building cul-de-sac for ever, endlessly talking to ourselves while managing the pain in unhealthy ways – something I know a lot about. For those who want to change the world, go forth and do your work – you don't have to be a paragon of healing to make a start, just like you don't need to be an expert to tell your story. But that parallel inner work cannot be postponed until after the revolution – the revolution begins within. Those who dismiss this as neo-liberal cliché are welcome to their beliefs. Beliefs that are too often as original as their ascribed identities and incomplete narratives will allow – don't I know it. Taking refuge from reality in our theories, frameworks and counterfactual fantasies may suffice for a time, but if our political activity is not earnestly performed, and serves mainly as a distraction from our internal dysfunction, then one day the weight of our nonsense will bring us to our knees. I once took pride in being equal parts informed and miserable. The world was something I thought I had all figured out. In truth, what I knew of life then you could write on the back of a cigarette packet – leaving space for all three volumes of *Das Kapital*. My simmering toil, that agitated emotional signature that passes so often for righteous anger, arose not always from a deep concern for the misfortunate I often hoped to portray. It just as often lay in the rage of waking every day to find the same misbehaving world, refusing to fit itself to my ideological straitjacket. In any event, true idealogues must usually learn painfully from their own mistakes (or see others pay for them) before accepting even basic facts about reality or themselves. Have it your way. But if I could be so bold as to offer just one suggestion before you move hastily on to the next digital battleground: take less direction on life from those who offer the same answer to every question. If your first act of healing is as simple as running as far as your knees will allow from that particular brand of blowhard, you will have made a most radical beginning. The closer we get to the truth of things, the better armed we are to cope, to respond and to decide. The sharper the faculties we may bring to bear should we decide to act upon the world around us. Living in the truth every minute of the day is an impossible feat; learning to notice when we're living out of it provides a vital waypoint – our bodies have ways of telling us what the next right action is, and our job is to act when it speaks to us. Recovery occurs when we begin discerning between the voice of the terrified inner child and the reassuring whisper of a more loving, wiser part of ourselves. Only then may we begin telling a new story. Starting the redraft can feel daunting. Much like putting off a diet, we sense it will be hard, so we avoid it – easier to announce we're going on a diet than to stick to it once we start. For many of us living with trauma, the difficulty we're avoiding through storytelling, whatever form that story takes, is precisely what we must lean into. And the first truth we may have to confront is possibly also the hardest: the stories we tell ourselves aren't always as true as we think.

Now, as my parting gift, I will leave you with a last-minute revelation that occurred to me only hours before completing this body of work. I was rereading the manuscript before filing to my long-suffering publisher, when an old memory I'd locked away became suddenly available to me. A memory of the last time I saw my granny. My previous recollection was that there was no 'last time', save for the day she lay on her deathbed surrounded by my family while I remained at home drinking myself silly. I'd always written my decision to stay away off to my alcoholism, as this seemed to me a plausible on-brand explanation. But as is often the case, the truth reveals another story. My refusal to visit my granny on her deathbed had nothing to do with my drinking – I had already said my goodbye. And it was a farewell I knew I couldn't top a second time round, so I chose to keep my distance, certain she'd heard everything she needed to from her eldest grandchild. A week-or-so before she died, I visited her in hospital. She was, at this time, very sick and things weren't looking good, but she was up and about, and fully aware of what was going on. The night before what sadly became our final encounter in this world, I sat up late and wrote her a long letter – the kind of letter I'm sure she'd written many times in her storied life. In that letter, I told her how much she meant to me, how grateful I was to have known her, that she was my hero and that I would miss her safe and reassuring presence terribly. I poured my breaking heart into every word, conveying to her – as fully as my grasp of language would permit – what a truly wonderful woman she was. I remember giving her a cuddle before handing her the envelope and (I think) kissing her forehead. What she said to me, I do not recall. But as I walked away, glancing back one more time to give her a wave from the door of the ward, she smiled back like she knew something I didn't. If memory serves, the letter ended with an excerpt from the poem 'Sonnet XVII' by Pablo Neruda – the same poem I had my sister-in-law read at my wedding almost two decades after my granny passed away. I reproduce some of it for you now:

I love you without knowing how, or when, or from where,
I love you straightforwardly, without complexities or pride;
So I love you because I know no other way
Than this: where I does not exist, nor you,
So close that your hand on my chest is my hand,
So close that your eyes close as I fall asleep.

That was the last time I saw my granny – and that is a beautiful story which, for some reason, I forgot to tell myself. Perhaps such an act of sensitivity didn't fit my cultivated, tortured identity back then. Perhaps I required a more elaborate rendering so as to retain the continuity of tragedy then so central to my narrative. Or perhaps I just wasn't ready to experience the fullness of my grief – easier to talk about it than truly feel. In any event, I have no doubt, as life unfolds in the years ahead, whatever highs and lows lie in store, that many other truths will be revealed. Truths that may require further redrafting of my story, which I'll happily undertake – though I'll mercifully spare the rest of you. I wish you well in life and I thank you for the time you have given to my book – good luck with your story. As for my recent difficulties, and the trauma I've been living with this past year, it had been my intention to disclose the details. Sadly, for the ambulance-chasing voyeurs among you, I've since thought better of the impulse – that story is nobody's business but mine.

Acknowledgements

The prospect of writing a book about trauma took time to get my head around. I knew it would have to be a deeply personal affair and that with such an undertaking, I'd face many of the challenges outlined in this volume. I was also acutely aware that while trauma has always been a theme of my work, it wasn't a concept I had ever tackled head-on. It took time to summon the necessary confidence to pitch the initial idea and even more time to persuade others I was up to the task – not just as a writer but as a person who's been very open about their vulnerability. My thanks must therefore go to Andrew at Ebury, for being curious enough to embark on a journey of discovery with me despite his early reservations. I was given the time and space to let my ideas percolate. The concept for the book crystallised when I realised I didn't have to write an onerous tome filled with citations and arse-covering caveats and that the narrower theme of stories was both appropriate to my level of expertise and provided a much clearer through-line. The trickier part came when trying to braid traditional non-fiction with meta-commentary – a high-risk approach with many potential pitfalls. When I was finally in a position to pitch more clearly my ambition for the book, Andrew threw all of his weight behind it and my intermittent confidence grew as a result. While I am in no way privy to the complexities of the literary world, I'm not so naive that the pressures placed on publishers by various commercial and cultural imperatives have escaped my notice. Andrew became a real advocate for me when, I suspect, there may have been confusion or even doubt in some departments as to why I would switch lanes from a moderately successful formula of writing about social inequality, to one dealing with trauma and, ultimately, mental health. Andrew's feedback has always been incredibly insightful, but the considerate and sensitive manner in which it is so generously given has become an invaluable part of my creative process. Thank you, Andrew. I hope we get to work together again in future. I would also like to express gratitude for the support of everyone at Ebury who contributed to this work whether through editing, cover design or promotion. Cameron did me countless solids during the mercifully brief but no less thorough editorial process. Needless to say, many a darling was slaughtered. Marta and David then cast fresh eyes over the manuscript, helping me clarify certain arguments and reconcile inadvertent contradictions. There are so many glaring issues with books before they are published, that we writers simply cannot or will not see. I have my share of bad habits. Chief among them, a tendency to begin the work before I have a solid plan in place for structure. Andrew suggested I frame each chapter as a question I should then attempt to answer – very simple but also revelatory for me. This process presented me with an opportunity to grow as a writer and accept help and guidance on weaker areas, which made completing the book a far more pleasurable experience. I also dialled down the writing style ever-so-slightly, aware I can over-complicate matters at times. The aim here was simply to be understood rather than demonstrate prowess on the vocabulary front. The real art in editing lies not in the notes given to you, but in how they're given. The sensitivity an editor demonstrates, particularly with a work as personal as this, can be the difference between a painful edit and a pleasant one. On this occasion, with the book being on the shorter side, I was able to really get my teeth into the text, bringing to bear all I had learned from two very different editorial experiences with each of my previous books. You don't always get to meet every person rooting for the success of a given project from behind the scenes, but the immense efforts of others on my behalf is, I am happy to say, pleasantly palpable. I would like to offer my heartfelt thanks to Leeann White and Lisa Selby for their powerful and brave contributions. I am especially grateful to them for their openness and trust with respect to their stories, which I hope I've done justice on these pages. To James, Marlyn, Miriam and Suzanne – thank you for asking me to think harder by simply making your values visible. The period over which we engaged with these issues together was a real turning point in this book's overall trajectory. I fear had it not been for our discussions, I may have produced something far less charitable in nature. While mean-spirited books have no trouble finding dedicated audiences, such a work would have been a poor reflection of my values. May I also thank my agent Vivienne, and her wonderful, dedicated team, Cheryl and Carol, who each, in their own way, helped me keep the wheels from falling off my whole operation when life began coming at me fast in February 2024. I'll never forget the day I was due to travel to London to interview Bernie Sanders, and instead found myself at Accident and Emergency in my home town with what I believed was a heart attack. My body was sounding an alarm I was not yet ready to hear. As my health deteriorated, Vivienne and the rest of the team provided vital support both moral and practical that afforded me the vital breathing space necessary to get my bearings. I am so grateful to you all for going over and above for me time and time again. To my family, immediate and remote, thank you for accepting that this is what I want to do with my life. I know some of you may worry about me from time to time, and that my profile likely brings the occasional wolf to your door. All I can say is that while my form of public storytelling is not without its downsides, the positive impact it has when it's done right makes the sacrifices worth it. At least, this is my hope. This book marks the end of a particular phase in my life and career. A period of immense adjustment. The catalyst for embarking on this crazy journey was surely the news in 2015 that I was to become a father the following year. From the day my son Daniel was born, my life took on a new and less self-centred meaning. My daughter Lily arrived in 2018, when the tornado of my career began forming. It's been one hell of a ride for us, let me tell you. Admittedly, some of it was bumpier than I would have liked and much of that early turbulence was generated by yours truly. I'm sorry for the relapses and the dishonesty and selfishness that always precede them. I regret the pain I put you all through when I took leave of my better nature and lifted a drink in anger, despair or self-pity. While acting from my own sense of victimhood, I retransmitted my pain onto each of you in different ways. While I cannot change the past, I know that repair begins with acknowledgement of the burdens my behaviour placed on you. I pray I have since demonstrated through action my commitment to recovery, and our family, to the extent that verbal reassurances are no longer necessary. Please also know that there is no statute of limitations on my willingness to make amends, or even just hear the story of how I hurt you or let you down. That door will always be open, and you may walk through any time, any place – I'll always be around ready to listen. While the spoils of success have allowed me to provide certain things that may otherwise have eluded us as a young family until much later – financial security, work–life balance, a fairer division of domestic labour between my wife and I – I also know there were times when I became consumed by extra-familial commitments. I only hope that over these past couple of years, having learned some hard lessons, that my presence at home has been felt that bit more. The seemingly mundane daily routines that comprise raising children have become cornerstones of my life, anchoring me in times of adversity and stress. For many years, I never felt like I had a home. I possessed no real homing instinct. A dwelling was simply a place to sleep. Today, I relish my return to our cluttered little nest whenever life demands I go travelling. Wherever the wind takes me, from Birmingham to Barcelona, I'll always be in a rush to get back to you.

Darren McGarvey grew up in Pollok, Glasgow. He is a writer, hip-hop artist, broadcaster and campaigner. His bestselling and acclaimed first book *Poverty Safari* was awarded the Orwell Prize for political writing in 2018.